Mechanical and Thermal Properties of Engineering Thermoplastics Composites

nova
science publishers

www.novapublishers.com

Materials Science and Technologies Series

Thermal Modeling Reimagined: Recent Research Trends and Applications
Pankaj Thakur, PhD
and Jatinder Kaur, PhD (Editors)
ISBN: 979-8-89530-463-1
eBook ISBN: 979-8-89530-645-1

A Closer Look at Ion Transport
Jun Mei, PhD (Editor)
ISBN: 979-8-89530-418-1
eBook ISBN: 979-8-89530-533-1

Properties and Applications of Butadiene Rubbers
Soney C. George (Editor)
ISBN: 979-8-89113-969-5
eBook ISBN: 979-8-89530-062-6

Physics and Mechanics of New Materials and Their Applications – 2023
Ivan A. Parinov, DrSc, I. I.,
Shun-Hsyung Chang, PhD
and Erni Puspanantasari Putri, PhD
(Editors)
eBook ISBN: 979-8-89113-839-1

Properties and Uses of Metallic Glass
Shiv Prakash Singh, PhD (Editor)
ISBN: 979-8-89113-569-7
eBook ISBN: 979-8-89113-631-1

Innovations and Applications of Fiber-Reinforced Concrete
S.R.R. Teja Prathipati, PhD,
Yeswanth Paluri, PhD,
Hanuma Kasanagi, PhD
and Vijay Kunamineni, PhD
(Editors)
ISBN: 979-8-89113-465-2
eBook ISBN: 979-8-89113-529-1

New Research on Thermal Stresses
Pankaj Thakur, PhD,
Jatinder Kaur, PhD
and Pushap Lata Sharma (Editors)
ISBN: 979-8-89113-485-0
eBook ISBN: 979-8-89113-555-0

More information about this series can be found at
https://novapublishers.com/product-category/series/materials-science-and-technologies/

Senthilkumar Krishnasamy
Thitinun Ungtrakul
Sabarish Radoor
Rajini Nagarajan
and Suchart Siengchin
Editors

Mechanical and Thermal Properties of Engineering Thermoplastics Composites

nova
science publishers
www.novapublishers.com

DOI: https://doi.org/10.52305/ERCI6604

NOTICE TO THE READER

Library of Congress Cataloging-in-Publication Data

ISBN: 979-8-89530-460-0 (Hardcover)
ISBN: 979-8-89530-682-6 (eBook)

Published by Nova Science Publishers, Inc. † New York

Contents

Preface

In recent years, the demand for high-performance materials that are lightweight, durable, and cost-effective has grown exponentially across various industries. Among the emerging solutions, thermoplastic composites have gained significant attention for their unique advantages over traditional materials and thermosetting composites. Their rapid processability, recyclability, environmental resistance, and lack of volatile emissions during fabrication position them as prime candidates for next-generation applications in automotive, aerospace, electronics, healthcare, and construction sectors.

This book explores the transformative potential of engineering thermoplastic composites, focusing on their mechanical and thermal behaviors when reinforced with synthetic and natural fibers. By exploring advanceD materials such as polyamide 66, polyamide 6, polyamide 11, polycarbonate, poly(phenylene ether), polysulfone, thermoplastic polyester, poly(phenylene sulfide), polyetherimide, thermoplastic polyketone, and polybenzimidazole, we aim to provide a comprehensive resource that highlights the superior characteristics and versatile applications of these composites.

We believe this book will serve as a catalyst for new research, foster technological advancements, and support the growing need for lightweight, high-strength materials in future industry developments.

We sincerely hope that readers will find this book to be an informative guide on the remarkable journey of engineering thermoplastic composites.

Acknowledgments

We would like to express our heartfelt gratitude to our institutes (PSG Institute of Technology and Applied Research, India; King Mongkut's University of Technology North Bangkok (KMUTNB); and Jeonbuk National University, Republic of Korea) for their constant support and encouragement throughout the development of this book. We are deeply thankful to Nova Science Publishers for their trust and guidance during the publication process. We also extend our appreciation to the academic and research communities whose valuable contributions and insights have greatly inspired this work. Without their support, this book would not have been possible.

Chapter 1

Polyamide-Based Composites for Automotive Applications

Nurjannah Salim[1,2,*]
Nurul Huda Abu Bakar[1,2]
Siti Maznah Kabeb[1,2]
and Nurlin Abu Samah[1,2]
[1]Faculty of Industrial Sciences and Technology, Universiti Malaysia Pahang Al-Sultan Abdullah, Lebuh Persiaran Tun Khalil Yaakob, Gambang, Kuantan, Pahang, Malaysia
[2]Advanced Intelligent Materials Centre, Universiti Malaysia Pahang Al-Sultan Abdullah, Lebuh Persiaran Tun Khalil Yaacob, Kuantan, Pahang, Malaysia

Abstract

Polyamide (PA), also referred to as nylon, is one of the most significant engineering thermoplastic materials. It is renowned for its outstanding mechanical and physical properties and has proved critical in the automotive sector. Specifically, its lightweight property allows for manufacturing automobiles that consume less fuel. This chapter comprehensively discusses the use of PA composite in manufacturing automotive components. It will cover the issues of polymer structure, reinforcement and filler, processing technique, and future direction of PA composite. Additionally, it will review the application of PA and its composites reinforced with fiberglass and carbon fiber to highlight their potential as alternative materials in the automotive industry. Previously, air intake manifolds for automobiles were constructed from metal. However, recently, these components are commonly produced from 30-35% glass-reinforced PA 6. Similarly, grades 6,6 and 4,6 are also gaining

* Corresponding Author's Email: njannah@umpsa.edu.my

In: Mechanical and Thermal Properties of Engineering …
Editors: Senthilkumar Krishnasamy, Thitinun Ungtrakul, Sabarish Radoor et al.
ISBN: 979-8-89530-460-0

popularity and have similarly been used in manifolds. In short, replacing metal with nylon in the manifolds decreases production expenses by up to 30%, lowers the part's weight by up to 50%, reduces system costs by integrating parts, and enhances fuel efficiency.

Keywords: polyamide, composite, thermoplastic, automotive

1. Introduction to Polyamide Polymers

Polyamide (PA) polymers, commonly known as nylons, are a diverse class of synthetic polymers characterized by their robustness, versatility, and widespread applications across various industries [1]. Developed initially in the 1930s by chemist Wallace Carothers, PA has since become indispensable in fields ranging from textiles to engineering materials [2].

PA is widely used as an industrial thermoplastic in automotive, aerospace, and consumer products due to its advanced mechanical properties, shock resistance, and constant structural stability [3]. Nylon 6, a specific PA, is commonly used in food packaging papers and containers due to its unique barrier properties and puncture resistance. In the electronics industry, the material is valued because of its electrical insulation capabilities and resistance to back action and arc formation. Meanwhile, in automotive applications, nylon is preferred when designing bearings, gears, bushings, and conveyor belts due to its durability and excellent performance [4].

1.1. Chemical Structure and Molecular Characteristics

There are three types of PA polymers, each with unique chemical structure and molecule behaviors: Nylon 6,6, Nylon 6, and Nylon 12.

1.1.1. Nylon 6,6 (Polyhexamethylene Adipamide)

The monomer units of Nylon 6,6 are composed of Hexamethylenediamine (HMDA) and adipic acid, as illustrated in Figure 1. The production of Nylon 6,6 uses a polymerization process called polycondensation. This reaction is responsible for the material's high tensile strength and toughness, as well as excellent thermic and synthetic endurance. Moreover, Nylon 6,6 has good dimensional stability, which makes it suitable for various applications, including the plastics industries, thread, and synthetic yard goods [5, 6].

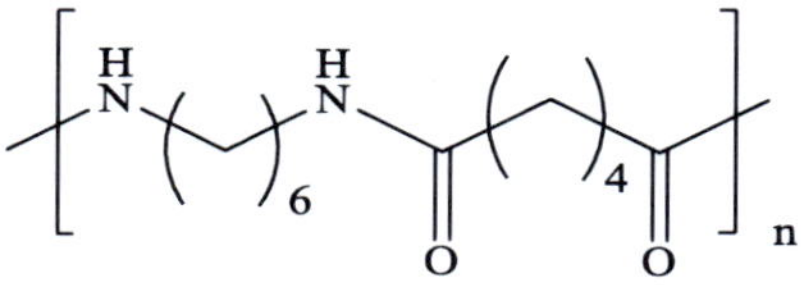

Figure 1. The structure of Nylon 6,6.

1.1.2. Nylon 6 (Polycaprolactam)

The monomer groups of nylon 6, through the polymerization process, called ring-opening polymerization, form ε-caprolactam, as depicted in Figure 2. Generally, Nylon 6 has excellent rigidity and flexibility, and it is highly resistant to abrasion and shock due to its high melting point. Due to these qualities, the material has been used in a variety of applications, such as fibers, textiles, and industrial plastics [7].

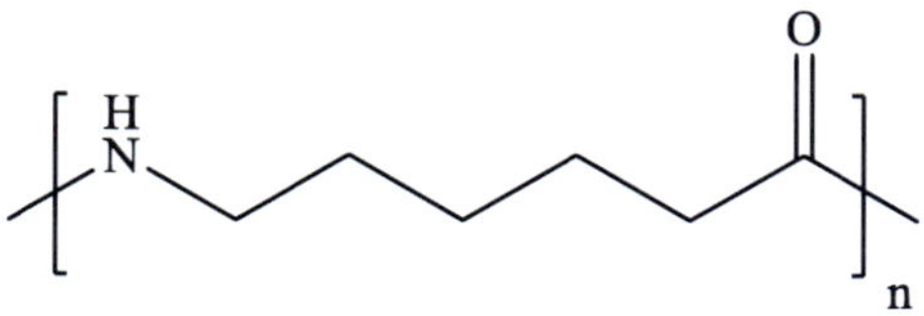

Figure 2. The structure of Nylon 6.

1.1.3. Nylon 12 (Polydodecamethylene Adipamide)

The monomer units comprising Nylon 12 are Dodecamethylenediamine (DDMA) and adipic acid, as shown in Figure 3. Similar to Nylon 6,6, the polymerization process to fabricate Nylon 12 also involves polycondensation. In general, Nylon 12 has good conversion and impact resistance; however, it is less absorbent. Due to its excellent chemical resistance, it is used in various industries, such as automotive fuel lines, airlines, and oil pipelines [3]. In conclusion, the variations in PA polymers determine their chemical structure and molecular attributes, resulting in a wide range of applications across many industries.

Figure 3. The structure of Nylon 12.

1.2. Classification of Polyamides

PA, or nylon, represents a huge class of materials with different chemical compositions, properties, and applications. They can be classified according to the chemical composition, polymerization method, or application, as displayed in Figure 4.

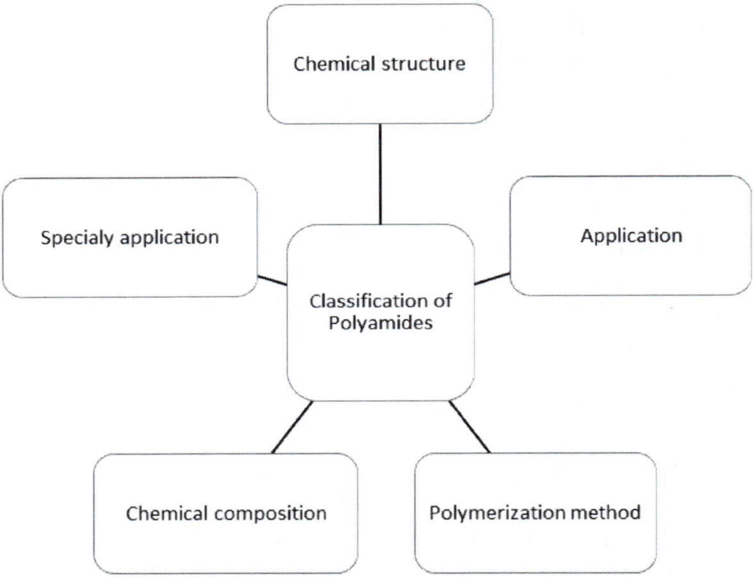

Figure 4. Classification of Polyamides.

1.2.1. Chemical Structure

In aliphatic PA, its backbone structure consists of aliphatic chains. Some examples of aliphatic PA include nylon 6, nylon 6,6, nylon 6,10, and nylon 12. In comparison, the backbone structure of aromatic PA is mostly made of aromatic groups that give it distinct properties, such as high thermal stability and mechanical strength. Some examples of aromatic PA include Aramid fibers like Kevlar and Nomex.

1.2.2. Method of Polymerization

In polymerization reactions such as polycondensation of PA, the formation of PA occurs through the condensation reaction between a dicarboxylic acid and a diamine (e.g., Nylon 6, Nylon 6,6, and Nylon 12). For the polymerization of lactams, the PA is produced through the ring-opening polymerization of lactams, which are cyclic amides (e.g., Nylon 6 and Nylon 11).

1.2.3. Chemical Composition

Polymer is commonly a material composed of different monomer units. However, when a polymer is made of repeating units of the same monomer, such as Nylon 6 (polycaprolactam), it is called a homopolymer. For example, copolyamides (or Alloys) are formed by copolymerizing two or more different monomers, such as Nylon 6,6 (polyhexamethylene adipamide) containing hexamethylenediamine and adipic acid. In conclusion, PA exhibits a wide range of properties and applications, making them essential materials in various industries, including textiles, automotive, aerospace, electronics, and consumer goods.

2. Polyamide Composites in the Automotive Industry

PAs are popular for their diverse qualities and are widely used in different areas of automotive manufacturing. Figure 5 shows an example of a PA composite that has been used in automotive applications.

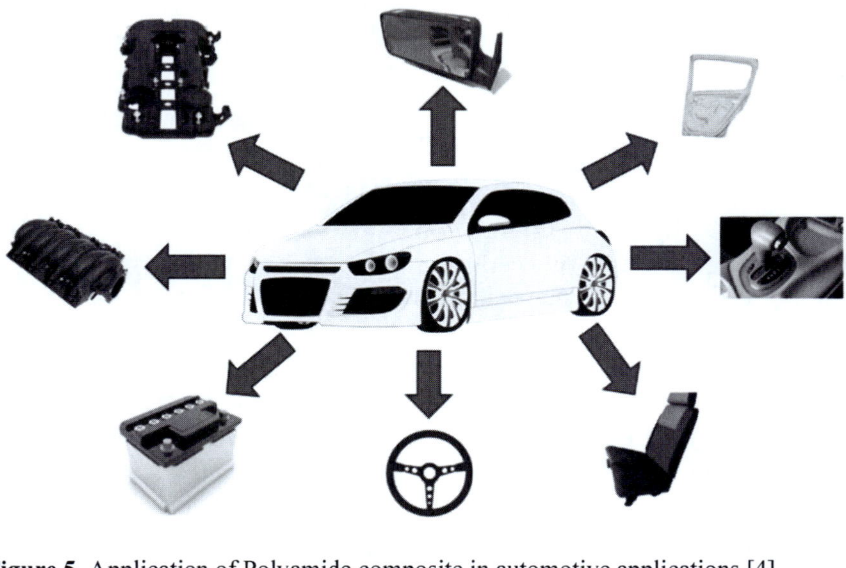

Figure 5. Application of Polyamide composite in automotive applications [4].

2.1. Engine Components

2.1.1. Intake Manifolds

To deliver the mixture of air and gasoline to the engine's cylinders, an intake manifold is necessary. In the current market, many intake manifolds are produced from PA composite reinforced with glass fiber, carbon fiber, mineral, and natural fiber. The uses of glass fiber-reinforced PA are preferable to other materials due to their extensive properties, such as high strength, good dimensional stability, and resistance to elevated temperature. Compared with traditional metal intake manifolds, PA composites are more suitable for engine components because they are lightweight and durable for automotive applications [5].

2.1.2. Valve Covers

The position of valve covers in the engine is around the camshafts and valves. Its function is to seal the cylinder head to prevent the engine oil from leaking. Historically, valve covers were produced from metal, but nowadays, PA composite is rapidly gaining popularity among consumers as a great alternative. This composite, reinforced with glass fiber, offers high strength, thermal resistance, and chemical resistance to extreme conditions in the engine area. Furthermore, as a composite material, PA composite has a low density, meaning that it is lightweight and can help increase fuel efficiency [6].

2.1.3. Oil Pans

The function of oil pans is to store the engine's lubricating oil. Lubricating is used to control the engine temperatures and to facilitate the oil circulation in the engine. Traditionally, oil pans are produced from metal. However, with the invention of PA composite using glass fiber reinforcement, oil pans that are lighter and more resilient to oil and temperature changes have been developed. In brief, PA composite is preferred because of its low density and ability to reduce emissions and increase fuel efficiency [7].

2.2. Interior Components

PAs composite can improve the attractiveness, functionality, and comfortability of automotive interiors. Examples of automotive interiors that use PA composite are dashboard panels, door panels, and seat structures [8]. In the composite manufacturing industry, PA composites are molded into various textures and colors, allowing designers to create visually appealing and ergonomic interior spaces. On top of that, since the automotive interior components designed with this material are lightweight and have excellent dimensional stability, this ensures a precise fit and finishing [6].

2.3. Exterior Components

Examples of PA composites that have been used as exterior components are wheel arch liners, bumpers, and grilles. These components are used to withstand collisions and maintain automotive structural strength. As such, it requires a material that can absorb impact strength, possesses high mechanical strength, and has the ability to withstand uncertain weather conditions, like PA composites. Additionally, the materials are suitable for exterior components because they have good resistance to UV radiation, making them ideal for outdoor applications [9].

2.4. Under-the-Hood Components

Internal components in automotive face uncertain conditions like expanded temperatures, chemical exposure, and mechanical strain. Therefore, since PA composite excels in diverse environments, it becomes the preferred material for several packages within the engine compartment. Among these parts, engine mounts, cooling device additives, air consumption ducts, and fluid reservoirs are commonly created from PA composite. Based on a conducted analysis, the use of this composite improves the sturdiness of those parts by withstanding heat cycling, chemical publicity, and mechanical loads while preserving their structure integrity through the years [5].

2.5. Electrical and Electronic Components

The rise in car electrification has caused an increased demand for substances that can satisfy the precise needs of electrical and digital additives. PA is considered the best for this application because of its superb electric insulation, thermal balance, and moisture resistance compared to other materials. Today, it is utilized in wiring harnesses, connectors, sensor housings, and manipulation of modules to ensure reliable performance in automotive electrical systems [10].

2.6. Structural Reinforcements

PAa are used both independently and as structural reinforcements in composite substances. For instance, PA/glass fiber and PA/carbon fiber composites can enhance the electricity, stiffness, and effect resistance of several automobile components. Notably, the material is frequently utilized in structural reinforcements for chassis components, suspension systems, and safety capabilities like crash-packing containers and aspect-effect beams [6, 7].

2.7. Fluid Handling Systems

PAs are typically utilized in automotive fluid control systems because of their potential to withstand lubricants, fuels, and car fluids. Some common examples of its usage are in manufacturing fuel traces, brake strains, and coolant hoses. This material is an exceptional choice for fluids dealing with applications because it can resist automotive fluids and preserve their integrity under strain and temperature variations [11].

2.8. Acoustic and NVH (Noise, Vibration, and Harshness) Control

PAs have natural dampening features that make them suitable materials for coping with noise, vibration, and harshness (NVH) within the vehicle cabin. In this aspect, PA-primarily based substances are applied to produce acoustic insulation panels, vibration damping pads, and engine bay coverings to decrease noise stages and beautify passenger consolation. Additionally, the composites can be customized to satisfy particular acoustic performance wishes, allowing manufacturers to adjust NVH characteristics according to purchaser demands and needs [12].

3. Polyamide Composites for Structural Automotive Components

3.1. Introduction to Polyamide as Automotive Components

The automotive industry is a global powerhouse with many sectors, including motor vehicle design, development, production, and sales. In this regard, the industry is considered a wide-ranging network of manufacturers, suppliers, and supplementary services, which jointly contribute to the growth of innovation and technology. In recent years, tremendous efforts have been made to improve the functionality and performance of the automotive components. Specifically, the vehicle body components, such as the hood, fenders, doors, bumper beams, and trunk lid, which constitute the vehicle's external framework, are of primary interest. On the other hand, seats, dashboards, steering wheels, and trim panels are among the internal components that have also been given immediate attention due to their critical function in enhancing passenger comfort, ergonomics, and overall visual appeal.

Regardless of the growth and development of vehicles, the automotive industry faces several challenges, especially in designing and manufacturing structural components. The key challenges include (i) the reduction of the overall weight of the vehicles by the implementation of lightweight materials for fuel efficiency, (ii) the urge to adopt environmentally friendly and recyclable materials to reduce the growing concern of environmental issues, (iii) ensuring the components are protected from detrimental effects of corrosion when subjected to harsh environmental factors, such as moisture and corrosive road salts, and (iv) pursuing cost optimization to curtail production expenses without sacrificing the quality and performance standard.

Numerous strategies have been employed to address the challenges mentioned above without compromising vehicle safety and performance. One strategy embraces the application of semicrystalline thermoplastic polymers, such as Polypropylene (PP) and PA, as a replacement for traditional steel alloys [13]. Table 1 lists a few common thermoplastics polymers with their respective applications in automotive components. Semicrystalline polymers exhibit stronger chemical resistance, higher coefficient of thermal expansion, and lower mold shrinkage than amorphous polymers [14]. It is also interesting to note that the level of crystallinity in the semicrystalline polymers could affect the yield strength, tensile strength, and chemical resistance.

Nevertheless, the increment in the crystallinity may reduce the elongation properties and resistance to crack propagation.

Table 1. Semicrystalline thermoplastic polymers used in automobiles

Polymer	Acronym	Examples of automotive application
Polypropylene	PP	Bumper beam, instrument panel, interior consoles, heating and air conditioning ducts
Polybutylene terephthalate	PBT	Mirror or sensor housings, fuse boxes, and electrical connectors
High-density polyethylene	HDPE	Ducts, fuel tank, and fluid bottle for windshield washer
Polyamide-6 (nylon-6)	PA-6	Cooling fan blades and engine throttle body
Polyamide-4,6 (nylon-4,6)	PA-4,6	Chain tensioners, electrical connectors, charge-air coolers
Polyamide-6,6 (nylon-6,6)	PA-6,6	Gears, cooling fan blades, resonators, and switches
Polylactic acid	PLA	Door trims, floor mats, tires, and consoles

Thermoplastic polymers are frequently incorporated with mineral fillers or fibers to improve the abovementioned properties. Various additives such as fire retardants, plasticizers, antioxidants, and colorants may be added to modify the physical properties of the polymer composites. Previous studies reported that polymer blends from two or more compatible thermoplastic polymers can be melt-blended to achieve the desired characteristics for automotive components [15]. For instance, PBT can be blended with polycarbonate (amorphous polymer) to be used as bumper beams [9] or PA-6,6 blended polyphenylene oxide (PPO) to improve heat resistance for outer body panels.

Overall, thermoplastic polymer composites are deemed a compelling opportunity than heavier metal components due to their intrinsic lightweight capabilities. This characteristic allows automakers to efficiently develop sustainable transportation that complies with environmental requirements and meets consumer expectations. The benefit of using thermoplastic matrix composites consists of reduced processing instances, greater weldability, and potential damage resistance. Most importantly, thermoplastic polymer composites provide recyclability opportunities through a pelletizing operation, which means that they can be further processed into new products. This, in turn, contributes to a greater sustainable and round usage cycle.

Among the diverse thermoplastics used in motors, PA has become the most widely used material because of its relatively low density than metals. In accordance with the industry's recognition of sustainability and environmental responsibility, utilizing lighter components, which include PA, can enhance gasoline efficiency and mitigate emissions. PA is normally characterized by the presence of amide groups (-CO-NH-) in their molecular chains. They are derived from the condensation polymerization reaction between a diamine and a dicarboxylic acid or their derivatives. The PA family includes a wide range of materials that are characterized systematically according to their distinct characteristics. This inherent variety allows for customizing car additives to meet customers' particular favored qualities.

Specifically, PA-6 (nylon 6) is produced by the polymerization of caprolactam (cyclic monomer), PA-4,6 (Nylon 4,6) is formed through polycondensation of adipic acid and 1,4-diaminobutane, and PA-6,6 (Nylon 6,6) is the product of condensation reaction between hexamethylenediamine (HDMA) and adipic acid. Meanwhile, bio-based PAs like PA-eleven and PA-6,10 are regarded as a sustainable opportunity for petroleum-based PA. Interestingly, the 100% bio-based PA (PA-11) exhibits a low melting temperature at 200 °C with remarkable mechanical performance that is comparable to other petroleum-based PAs. In addition, bio-based PA, i.e., PA-6,10, could be used to fabricate automotive components such as car interior door handles, which were previously made from glass fiber (GF) reinforced polypropylene (PP) [16].

Despite its hygroscopic properties due to the presence of oxygenated groups in its polymer chain, PA composites offer high advantages compared to other conventional thermoplastic polymers. Also, the processing method for PA-based composite is less complex, as it uses lower temperatures, cheaper raw materials cost, and compatible chemical temperature (40-50°C). Due to this versatility, PA has been utilized in safety components such as airbags to provide crucial protection during collisions [17]. Furthermore, PA is extensively employed in manufacturing protective enclosures and casings for car engines, as well as for the construction of vehicle body structures (exterior) and interior components. By utilizing the inherent lightweight nature of PA materials, automotive manufacturers have successfully reduced the overall mass of vehicles, optimizing both fuel efficiency and curbing fuel consumption rates.

3.2. Reinforcement and Fillers in Polyamide Composites

Current research endeavors are mostly directed toward integrating reinforcing agents into PAs. Compared to the original polymers, several physical properties can be improved by incorporating fillers into PA matrices, including heat deflection temperature (HDT), tensile strength, mechanical strength, stiffness, and impact resistance. Researchers have investigated reinforcing fillers, such as glass fibers [18], carbon fibers [19], mineral particles [20], or nanomaterials like graphene nanoplatelets (GNP) [21]. These enhancements are achieved by employing an effective PA matrix that interacts with the high-strength reinforcing fibers or particles. Besides that, researchers have investigated the integration of natural fibers, such as kenaf, cellulose, and biochar, with other substances like black carbon and chitosan (CS) into PAs [22]. According to Ozen et al. [23], the performance characteristics of natural fiber-reinforced polymer composites depend on the inherent qualities of the fibers used, their orientation within the matrix, and their concentration levels.

Surprisingly, reinforcing PAs with plant-based fibers has been found to enhance the mechanical strength of the polymers due to the hierarchical structure of the crystalline cellulose [24]. Baniasadi [25] demonstrated that the bio-based PAs of PA-11 and PA-12, which have a relatively low melting point, were strengthened by adding microcrystalline cellulose (MCC) that was grafted with octadecyl isocyanate. In addition, miscanthus-based biochar in PA-6 was proposed to improve the compatibility of the biocomposite. This enhancement may be attributed to the connection between the nylon backbone groups and the biocarbon, potentially facilitated by hydrogen bonding and polar-polar interactions [26]. The growing interest in incorporating plant-based reinforcement is to acquire the desirable characteristics of PAs as biodegradable and sustainable materials. Also, natural fibers are said to offer advantages over glass fibers since they are more ductile, less abrasive, and cost-effective. Moreover, this class of fiber has a significantly lower density of approximately 1.5 g/cm^3 compared to glass fibers that range from 2.5 to 2.8 g/cm^3, suggesting its reliability as a lightweight automotive component. Nevertheless, using natural fiber in the composite preparation processes may present a technical obstacle due to its limited heat stability.

As a result, glass fiber-reinforced PA remains a relevant filler in the automobile industry owing to its easy manufacturing process and advantageous mechanical characteristics. Incorporating glass fibers into PA composites imparts unique characteristics that cannot be obtained when individually using the fibers or the polymer. The mechanical behavior of the PA composite reinforced with fibers depends on many factors, including the bonding strength between the fibers and the PA matrix at the interface. Overall, the automotive industries require the use of low-density reinforcement to enhance the thermoplastic strength, modulus, and density beyond the base materials. Table 2 summarizes the examples or reinforcement materials for better comparison.

3.3. Processing Techniques for Polyamide Composites

PAs demonstrate a wide range of processing capabilities, and based on the selection of the processing method, unique geometric characteristics of the product can be achieved. Among the prominent approaches, conventional injection molding is widely recognized and employed in the automobile industry. Nevertheless, its strict requirement for precise dimensions and lightweight properties has spurred researchers to find other alternatives, such as microcellular injection molding (MIM). The MIM method is known for its capacity to produce complex, dense components with a cellular core and a solid outer layer [27]. PA-6,6 reinforced glass fiber is one example of the polymer prepared using the MIM. In this method, processing parameters, such as gas injection pressure and injection temperature, must be carefully controlled [28]. Based on a previous study, the most apparent effects of using this approach are on the polymer's morphology, density, thermal characteristics, and mechanical properties. Nowadays, injection molding is an extensively employed method because it enables manufacturers to acquire finished products with precise shapes and properties. The method can be classified into different categories: co-injection moulding, fluid-assisted injection moulding (gas-assisted or water-assisted), and in-mold decoration (IMD) injection moulding.

In another study, a three-stage extrusion method was investigated for its suitability in attaining a uniform distribution of nano clay within the PA-12 composite materials [29]. Additionally, Shamsuri [30] employed compression molding, a method consisting of three steps: pre-heating, applying compression to shape the polymer into the required form, and then cooling it down to achieve the desired permanent shape. Alternatively, composites consisting of PA-6 and leather waste were produced by blending these constituents using a Brabender apparatus at a set temperature of 230°C. Meanwhile, Ramezani-Dana [31] prepared the PA-6,6 composites with glass fiber fabrics using the thermo compression method with 600 kN capacity hydraulic vertical press at a maximum temperature of 400° C.

Graphene nanoplatelets (GNP) at various weight percentages are mix with PA-6,10 matrix incorporated using melt compounding with a twin-screw extruder to improve thermal and mechanical properties for automotive application [32]. In addition, modified carbon fibers (CF) reinforced PA-6 was prepared by submerging CF in a solution of PA6 in formic acid. The CF undergoes activation at a temperature of 400 °C in a nitrogen atmosphere for 4 hours to eliminate the sizing agents and improve the interfacial adhesion between the fiber and matrix in CF/PA6 composites [19]. On the other hand, PA-6/calcium carbonate was synthesized using injection molding, and it is worth mentioning that, in this study, both mechanical performance and processability were achieved at a rate as low as 1.5 wt.% of nano-calcium carbonate [20]. Meanwhile, the melt processing method was used to prepare the PA-4,10/biocarbon composite, where the corn cob was pyrolyzed at temperatures ranging from 400°C to 900°C [33]. In summary, various processing techniques can be utilized to fabricate PA-based composites suitable for automotive applications, as outlined in Table 2.

Table 2. Examples of reinforced materials and processing methods for PA

Reinforced materials	PA	Processing method
Glass fibers	PA-6,6	microcellular injection molding (MIM)
Glass fibers	PA-6,6	Thermocompression
Carbon fibers	PA-6	Surface coating treatment
graphene nanoplatelets	PA-6,10	melt compounding with a twin-screw extruder
biocarbon	PA-4,10	Pyrolysis, melt processing
calcium carbonate	PA-6,6	Injection molding
nano clay	PA-12	three-stage extrusion
Leather waste	PA-6	blending

Automotive manufacturers have effectively utilized the inherent lightweight properties of PA to decrease the total weight of vehicles, resulting in enhanced fuel efficiency and reduced fuel consumption rates. In addition to being lightweight, PA-based composites are also crash-worthiness, generally corrosion-resistant, and cost-effective. Thus, the material can be deemed in line with the industry's collective endeavors to advance sustainability and environmental awareness for more environmentally friendly transportation alternatives.

4. Emerging Trends and Future Directions in Polyamide Composites for Automotive Applications

PA composites are revolutionizing the automotive industry by offering lightweight construction, robust mechanical properties, and design versatility [34]. As the automotive landscape evolves towards sustainability and better performance, these composites stand at the forefront of meeting regulatory requirements and consumer preferences for eco-friendly vehicles [35, 36]. The importance of lightweight materials has increased dramatically, with PA composite becoming the preferred material because it is stronger, stiffer, and lighter than conventional materials [4, 37]. Their customizability allows for complex designs and sophisticated geometries that enable automobile manufacturers to increase performance while meeting safety standards [38, 39]. Today, PA composites are widely used in many automotive applications as the primary material for structural components, interior trim, and under-the-hood applications. Their use greatly improves the vehicle's power, fuel efficiency, and overall durability.

4.1. Sustainability and Eco-Friendly Polyamide Composites

PA blend composites have attracted attention in the automotive industry for their performance and ability to support sustainability efforts. Recently, the emerging trends in the development of PA focus on sustainability and greater environmental responsibility, which aligns with the industry's increasing emphasis on environmental responsibility. Owing to this latest shift in interest, researchers have also begun to emphasize a multifaceted approach to improving PA composites in the field, balancing performance with durability and functionality.

Today, the trend is towards sustainable automotive materials through the incorporation of bio-based materials, such as natural fibers or agricultural by-products, into PA formulations. Many have proved that this approach can significantly improve the mechanical properties of polymers, suggesting that its efficiency is comparable to traditional fillers such as glass fibers [40-42]. Cellulose fibers, in particular, provide better fuel efficiency in PA composites than glass or carbon fibers, with the additional benefits of being cost-effective and suitable for various automotive applications [21, 43].

DENSO, in collaboration with Dupont, introduced a cooling water box made of bio-based PA610, the first bio-based technical component in engine compartments, known for its durability in high temperatures and corrosive environments [44]. Other car brands, including Peugeot and Volkswagen, have also utilized components similar to Rilsan HT to replace metal hoses found under their car hoods. These changes demonstrate the growing adoption of plant-based PAs in automotive engineering [45]. Likewise, Royal DSM NV introduced EcoPaXX, a high heat-resistant castor oil-based PA that offers nearly equivalent performance to traditional petroleum-based materials, with over 50% biobased content and is typically used in automotive engine compartments [46].

The integration of nanotechnology into PA composites offers an alternative approach to improve their sustainability while meeting the rigorous performance demands of automotive applications [47]. This integration aligns with the broader concept of polymer nanocomposites (PNCs), where nanomaterials are strategically incorporated into polymer matrices to increase various performance attributes [48]. Understanding the role of nanofillers in PNCs means that researchers can now explore various potential applications and benefits in automotive engineering. This, in turn, will contribute to developing lightweight and durable vehicle components. For instance, nanofillers like nanoclays, carbon nanotubes (CNT), or graphene can bolster mechanical properties and thermal stability while reducing waste. Chatterjee et al. [16] investigated the influence of adding CNT and GNPs on PA 12 (PA12). The study revealed a remarkable increment of efficiency up to 80% in the toughness modulus of CNTs treated with a surfactant over pure PA12.

PNCs enhance technologies in areas such as automotive paints, catalytic converters, exhaust systems, transmissions, and engines by offering cost-effective, lightweight solutions. The material's mechanical properties are crucial for vehicle design because their weight reduction benefit can improve fuel economy and reduce pollution [49]. Automakers like BMW and Peugeot use PNCs to produce automobile components [48]. The addition of appropriate nanoparticle concentrations (even at 5 wt.%) significantly enhances mechanical properties, particularly the strength of a pure polymer. This improvement correlates with the degree of interaction between nanoparticles and the polymeric matrix, where a higher level of interaction leads to greater mechanical strength [48]. Nanofillers such as nano clay not only enhance mechanical properties but also improve thermal stability by acting as thermal insulators and mass transport barriers. This consequently raises the glass transition temperature (T_g) and crystallization enthalpy while also serving as nucleating agents that promote crystallization and increase crystalline melting temperature (T_m) [50].

In automotive applications, barrier properties and chemical resistance are crucial characteristics that ensure durability and performance against gas permeation and chemical exposure for fuel systems, exhaust systems, and coatings. [51] discovered that incorporating 5 wt% of montmorillonite (MMT) into PA12 significantly enhances its barrier properties, particularly in terms of corrosion protection, compared to PA12 alone. In a separate study, [52] found that subjecting PA6 or PA66-based long carbon fiber composites to temperatures exceeding 190°C for 20 minutes during the painting process does not degrade their mechanical properties.

Furthermore, advancements in processing techniques and recycling technologies are driving the development of sustainable PA composites [53]. By optimizing manufacturing processes to minimize energy consumption and waste generation, manufacturers can enhance the eco-friendliness of PA composite production. This includes optimizing the drying process for PA pellets, which is critical due to their water-absorbing nature. If the step is handled inappropriately, the drying conditions can significantly reduce the lifespan of manufactured parts. Generally, it is recommended to dry these pellets at temperatures between 70 and 90°C for up to 12 hours in an oven [54]. Furthermore, the development of processing techniques helps to address challenges such as post-processing shrinkage, which is common in PAs and particularly noticeable in injection molding processes. By improving processing techniques, manufacturers can better mitigate issues related to post-processing shrinkage and ultimately reduce the process costs [53].

Additionally, innovations in recycling methods enable the recovery and reuse of PA materials. This repetitive process, in the long term, extends their lifecycle and reduces environmental impact. PA-based materials, with their high recycling potential, are drawing attention from consumer industries due to the feasibility of producing quality materials for reinserting into the production process or use as raw materials in other sectors [4]. [55] discovered that despite reduced mechanical properties due to shorter fibers, the recycled material from vehicle end-of-life radiators, composed of PA66 reinforced with 35.7 wt. % short FG remains suitable as a replacement for automotive components. This effort is in accordance with stringent EU regulations, which require at least 85% by weight per vehicle for reuse and recyclability and at least 95% by weight per vehicle for reuse and recovery. [56] in their study demonstrated that there is no variation in strength and stiffness when utilizing recycled or virgin carbon fiber in the production of PA66 composite-based materials for automotive applications. However, the recycling of PA presents challenges, as difficulties in waste separation and cleaning, coupled with the necessity for meticulous management of repetitive melt processing, can lead to the degradation of PA properties [57].

Overall, the exploration of sustainability and eco-friendly PA composites lays the groundwork for advancements in smart and functional PA composites, showcasing a progressive approach to the development and application of PA materials in the automotive industry.

4.2. Smart and Functional Polyamide Composite

The development of PA composites in automotive applications has made significant progress as intelligent and functional materials. The combination of functional fillers and additives to match emerging trends allows the performance characteristics to be optimized by meeting specific requirements in automotive manufacturing and engineering [58]. In the automotive industry, there is a growing demand for materials that can not only improve structural integrity but also enhance performance. These products must have features such as lightweight, durable, shock-resistant construction, thermal stability, and chemical resistance to withstand the harsh conditions encountered in automobiles. Furthermore, increasing emphasis is being placed on integrating intelligent features into automotive components, such as self-healing capabilities, thermal management capabilities, and electrical protection, to enhance vehicle performance, safety, and comfort. Therefore, PA composites with customized properties play an important role in meeting the evolving needs of the automotive industry.

The addition of reactive additives such as phosphorus-based materials or halogen-free flame retardants increases the flame resistance of PA composites. Such improvement means that the material is now more compliant with safety standards. In particular, through this development, chemicals that are made of phosphorus can provide less toxic combustion gases while offering advantages such as greater molecular distinguishability that make them promising alternatives to halogen-free flame retardants [59, 60]. This integration of intelligent, active components not only increases productivity but is in line with the automotive industry's commitment to sustainability and safety. In 2020, [61] showed that the addition of DDP as a firewall flame retardation in PA6 by transesterification resulted in a V-0 UL94 test value and 29.5% LOI value, conforming to flame retardant requirements.

Furthermore, the introduction of conductive fillers opens new possibilities for electroconductive PA composites. Materials such as carbon nanotubes or metal nanoparticles provide electrical conductivity in PA matrices, enabling easy integration of electronic components in automotive systems. This advancement not only facilitates the incorporation of advanced technologies but also enhances electromagnetic shielding, ensuring optimal performance and reliability in electric and connected vehicles. Recently, [62] conducted the first broadband dielectric spectroscopy investigation into the temperature- and frequency-dependent electrical conductivity of PA6 composites featuring metal microparticles (Al, Fe, or Cu) and CNT. The research revealed a significant increase in electrical conductivity (nine to eleven orders of magnitude) in metal-CNT composite plates at 20 °C and 0.1 Hz frequency. The researcher credited the improvement to the presence of metal particles within the CNT conducting network.

Hybrid and multi-material systems represent another frontier in PA composite development, offering synergistic combinations to enhance performance characteristics. By integrating PA composites with materials such as carbon fiber or thermoplastic elastomers, automotive manufacturers can achieve unique blends of strength, flexibility, and lightweight construction. [63] successfully developed a nanocomposite comprising PA 6 and graphene oxide (PA6-GO). Based on the findings, notable enhancements of 21% in tensile modulus and 16% in yield stress at a filler loading of 0.5 wt% were made. Additionally, their findings revealed promising mechanical adhesion and exceptional shear strength (7.1 ± 1.0 MPa) when joining with aluminum alloy AA6061 in PA6-GO/AA6061 hybrid joints, underscoring its potential for lightweight engineering applications. In another study, a similar enhancement of lightweight automotive materials was done by improving the thermo-mechanical properties of wood fiber/carbon fiber bio PA hybrid composites. The study demonstrated that these modifications managed to increase the heat deflection temperature (HDT) value up to 127°C in the PA: WF10CF20 composite, an improvement attributed to the high thermal stabilities of synthetic carbon fiber [64]. Additionally, [54] found that PA6-based composites, with their favorable thermo-mechanical properties in addition to superior creep and fatigue resistance characteristics, can replace traditional automotive metal materials in manufacturing structural parts and components, such as engine compartments, bearings, and gears.

Hybrid systems also contribute to enhanced conductivity, as shown by [65], who observed that PA6/70 wt.% Al_2O_3 and PA6 with hybrid fillers at 75 wt.% displayed the highest thermal conductivity enhancement factors, up to 145% and 92%, respectively. These hybrid systems align with the industry's objectives of efficiency and environmental responsibility, as they offer enhanced functionality without compromising environmental considerations.

The integration of functional additives enhances performance attributes such as flame retardancy and electrical conductivity, while hybrid and multi-material systems provide synergistic combinations for enhanced strength, flexibility, and lightweight construction, aligning with industry objectives of efficiency and environmental responsibility. The burgeoning focus on sustainability in automotive engineering has sparked interest in bio-PAs, which will be explored in the subsequent section.

4.3. Bio-Polyamides in Automotive Applications

The utilization of natural fiber-reinforced PAs, especially those with cellulose fibers, presents challenges due to the processing temperatures that surpass the fibers' degradation limit. This issue highlights the imperative shift towards bio-Pas in order to align the materials with the sustainability goals outlined in section 4.1, as well as to meet the increasing demand for environmentally conscious materials within the automotive industry [66, 67]. Bio-PAs are gaining popularity in the industry for their recyclability, electrical resistance, ductility, and creep resistance. On top of these qualities, their capacity to improve fuel efficiency and weight reduction makes them more sought after by manufacturers. Today, biopolymers are utilized in various automotive components such as steering, doors, wheels, and electrical parts, as well as in aircraft interiors, waste systems, engine components, and exterior parts, suggesting their potential for wider adoption in aerospace industries with technological advancements [68].

Bio-PAs can be synthesized through various methods, including condensation reactions of di-amines with bio-derived di-carboxylic acids, condensation reactions of amino acids as bi-functional monomers, or ring-opening polymerization of lactams [69, 70]. Nonedible castor oil (CO), derived from the seeds of the castor plant (*Ricinus communis*), has been utilized as a renewable resource in various industrial applications, including for the production of bio-PAs [71]. CO contains a high proportion of ricinoleic acid ($C_{18}H_{34}O_3$), a monounsaturated fatty acid that imparts unique chemical properties. These properties include its ability to undergo esterification and amidation reactions in the presence of carboxylic acid groups. Moreover, the double bonds in ricinoleic acid also aid in hydrogenation and epoxidation reactions, while the hydroxyl groups allow for acetylation, alkoxylation, or dehydration reactions to take place [72, 73].

As a renewable and non-food resource, CO offers a sustainable alternative to petroleum-derived food sources for the production of bio-PAs and other polymeric materials. An example of this sustainable method is Arkema Bio-PA11, which is derived entirely from castor oil and exhibits reduced CO2 emissions compared to petroleum-based counterparts. Despite minor performance differences, the focus is on profitability for the environment [74].

There are different strategies to synthesize bio-PAs, each involving specific chemical modifications that reflect different mechanisms [70]. For example, 11-aminoundecanoic acid is used to synthesize the monomer for the AB-PA (PA11) using an adecenoic acid method. This approach is known to produce versatile bio-PAs and, since its introduction, has set the stage for exploring new applications in various industries. Accordingly, the study by [75] examines the impact performance of PA11 flax/basalt hybrid composite materials, highlighting the advantages of plasticized PA11. The findings revealed an improvement in glass transition and low-temperature hardness, as well as increased efficiency and delayed penetration of materials, all of which are indicative of its potential for semi-structural vehicle applications. However, the trace availability of the 11-aminoundecanoic acid monomer with ATO prevents the production of PA11 in developing countries, causing many countries to import PA11 from France. The issue has also raised concerns about counterfeit products such as PA1012 [76].

In contrast, AABB PAs, such as PA610 and PA1010, utilize sebasic acid (1,8-octanedicarboxylic acid) obtained via the sebasic acid pathway, a much more straightforward pathway in PA synthesis. Additionally, the diamine (AA) monomer for AABB PAs PA1010, PA1012, and PA10T is decamethylene diamine (DMDA or 1,10-diaminodecane). Since this monomer is derived from sebacic acid, it is considered to align with the sebacic acid route.

4.4. Future Directions and Innovations in Automotive Composite Materials

Future directions and innovations in PA automotive composite materials, such as glass fiber (PA-GF) or carbon fiber (PA-CF) composites, represent significant advancements in the automotive industry [77]. For example, the front-end carrier module in the Audi Q7 incorporates a high-performance PA-based composite material reinforced with either glass fibers or carbon fibers, chosen based on specific requirements such as weight savings and structural integrity. This structural component is noteworthy for its absence of sheet-metal reinforcement, even compared to other all-plastic front-end carriers. Furthermore, the high stiffness and strength of the PA allow for intricate, thin-walled designs.

In the near future, it is anticipated that the automotive industry will face challenges related to cost, product flexibility, and end-of-life disposal of PA blends. However, emerging materials and technologies such as better additives and smart materials are expected to help the industry overcome these challenges while also opening new possibilities for PA applications. As electrification and autonomous vehicle technologies continue to evolve, composite PAs are poised to play a key role in providing lightweight, efficient, and sustainable vehicle solutions. In short, by taking advantage of composite PAs' inherent characteristics and versatility, the automotive industry can drive innovation, increase productivity, and progress toward a more sustainable future.

Conclusion

In conclusion, polyamide-based composites are preferred by automotive manufacturers since this material produces comfortable and acoustic vehicles. Polyamide-based composites have outstanding vibration-damping qualities that lessen driver fatigue and effectively reduce cabin noise. These composites are being strategically incorporated by automotive manufacturers to provide a quieter and more comfortable ride without appreciably increasing vehicle weight. This feature is especially helpful in electric cars, where other background noise is amplified by the lack of engine noise.

Additionally, Polyamide-based composites show exceptional durability, enduring exposure to extreme environments like road salt, automotive fluids, and UV rays without deteriorating or losing their aesthetic appearance. Hence, vehicle components made from polyamide composite high quality, require less maintenance, and add to the vehicle's overall durability. Furthermore, Polyamide-based composite can be colored and textured during production, which reduces the need for extra painting and streamlines production while also benefiting the environment.

Polyamide-based composites play a key role in enabling novel features as automotive technology develops. These materials are used to create sturdy yet lightweight sensor and camera housings that improve driver safety. Additionally, they make it easier for radar and lidar systems to function effectively, guaranteeing the cars can see their environment. These adaptable materials will continue to stimulate innovative design and functional possibilities as automobiles become more electrified and connected. Additionally, some polyamide composites have excellent thermal management properties, which are essential for preserving ideal battery temperatures and extending battery life in electric vehicles.

References

[1] Zou, H., et al., A change of phase morphology in poly (p-phenylene sulfide)/polyamide 66 blends induced by adding multi-walled carbon nanotubes. *Polymer,* 2006. 47(22): p. 7821-7826.

[2] Viswanathan, A., Wallace Carothers: More than the inventor of Nylon and Neoprene. *World Patent Information,* 2010. 32(4): p. 300-305.

[3] Fernández, C. E., et al., Compared structure and morphology of nylon-12 and 10-polyurethane lamellar crystals. *Polymer,* 2011. 52(7): p. 1515-1522.

[4] Kondo, M. Y., et al., Recent advances in the use of Polyamide-based materials for the automotive industry. *Polímeros,* 2022. 32: p. e2022023.

[5] Behera, A., Additive Manufacturing Materials. *Advanced Materials: An Introduction to Modern Materials Science,* 2022: p. 667-700.

[6] Parin, F. N. and F. Demirci, Durability of polymer composite materials for high-temperature applications, in *Aging and Durability of FRP Composites and Nanocomposites.* 2024, Elsevier. p. 135-170.

[7] Njuguna, J., Z. Mouti, and K. Westwood, Toughening mechanisms for glass fiber-reinforced polyamide composites, in *Toughening Mechanisms in Composite Materials.* 2015, Elsevier. p. 211-232.

[8] Phadke, G. and D. Rawtani, Bioplastics as polymeric building blocks: Paving the way for greener and cleaner environment. *European Polymer Journal,* 2023. 199: p. 112453.

[9] Gül, Ç. and E. D. Kocak, Emerging Applications of Polymers for Automobile Industries. *Specialty Polymers,* 2023: p. 157-168.

[10] Turska, M., *High Voltage Components in Commercial Vehicles.* 2017.

[11] Tormos, B., et al., Compatibility Study of Polyamide (PA6) with Lubricant Bases for Electric Vehicle Applications. *Lubricants,* 2024. 12(2): p. 54.

[12] Harrass, K., S. Mauer, and R. Tsekov, Resistance of glass fibre reinforced polyamide 6.6 materials to automotive cooling fluids: An analytical method for lifetime prediction. *Polymer International,* 2022. 71(6): p. 724-733.

[13] Mohammadi, H., et al., Lightweight glass fiber-reinforced polymer composite for automotive bumper applications: A review. *Polymers,* 2022. 15(1): p. 193.

[14] Mallick, P. K., Thermoplastics and thermoplastic–matrix composites for lightweight automotive structures, in *Materials, design and manufacturing for lightweight vehicles.* 2021, Elsevier. p. 187-228.

[15] Simunec, D. P., et al., Facilitating the additive manufacture of high-performance polymers through polymer blending: A review. *European Polymer Journal,* 2023: p. 112553.

[16] Oliver-Ortega, H., et al., Research on the use of lignocellulosic fibers reinforced bio-polyamide 11 with composites for automotive parts: Car door handle case study. *Journal of cleaner production,* 2019. 226: p. 64-73.

[17] Ali, S., et al., Enhancing the properties of nylon 66 fabric coated with a combination of PVA and SiO_2 nanoparticles composite for vehicle airbag application. *Journal of Industrial Textiles,* 2022. 51(2_suppl): p. 3330S-3346S.

[18] Wei, X.-F., et al., Performance of glass fiber reinforced polyamide composites exposed to bioethanol fuel at high temperature. *npj Materials Degradation,* 2022. 6(1): p. 69.

[19] Hassan, E. A. M., et al., The effect of carbon fibers modification on the mechanical properties of polyamide composites for automobile applications. *Mechanics of Composite Materials,* 2022. 58(2): p. 261-270.

[20] Augusto, A., T., et al., Effect of injection molding conditions on the properties of polyamide 6/calcium carbonate nanocomposite. *Journal of Applied Polymer Science,* 2023. 140(30): p. e54087.

[21] Kiziltas, A., et al. Micro-and nanocellulose composites for automotive applications. in *Proceedings of the SPE Automotive Composites Conference and Exhibition (ACCE),* Novi, MI, USA. 2013.

[22] Shrivastava, A., M. Chakraborty, and A. K. Singh, Biocomposites with polyamide fibers (nylons and aramids), in *Advances in Biocomposites and their Applications.* 2024, Elsevier. p. 121-147.

[23] Ozen, E., et al., Natural fiber blend—nylon 6 composites. *Polymer Composites,* 2013. 34(4): p. 544-553.

[24] Kargarzadeh, H., et al., Recent developments in nanocellulose-based biodegradable polymers, thermoplastic polymers, and porous nanocomposites. *Progress in Polymer Science,* 2018. 87: p. 197-227.

[25] Baniasadi, H., et al., Sustainable composites of surface-modified cellulose with low–melting point polyamide. *Materials Today Chemistry,* 2021. 22: p. 100590.

[26] Mohanty, A., et al. *Sustainable composites from renewable biochar and engineering plastic.*

[27] Turng, L.-S., Special and emerging injection molding processes. *Journal of injection molding technology,* 2001. 5(3): p. 160.

[28] Volpe, V., et al., Lightweight high-performance polymer composite for automotive applications. *Polymers,* 2019. 11(2): p. 326.

[29] Stojšić, J., et al., A Study of the flexural properties of PA12/clay nanocomposites. *Polymers,* 2022. 14(3): p. 434.

[30] Shamsuri, A. A., Compression moulding technique for manufacturing biocomposite products. *International Journal of Applied,* 2015. 5(3).

[31] Ramezani-Dana, H., et al., Experimental investigation of the mechanical behavior of glass fiber/high fluidity polyamide-based composites for automotive market. *Journal of Reinforced Plastics and Composites,* 2021. 40(21-22): p. 827-844.

[32] Kiziltas, A., et al., Graphene nanoplatelet reinforcement for thermal and mechanical properties enhancement of bio-based polyamide 6, 10 nanocomposites for automotive applications. *Composites Part C: Open Access,* 2021. 6: p. 100177.

[33] Watt, E., et al., Biocomposites from biobased polyamide 4, 10 and waste corn cob based biocarbon. *Composites Part A: Applied Science and Manufacturing*, 2021. 145: p. 106340.

[34] Jubinville, D., *Process Engineering and Performance of Sustainable Polyamide 6 (PA6) Based Blends and Biocomposites.* 2018, University of Guelph.

[35] Heflin, D. G., *Hybrid Composite Materials and Manufacturing.* 2022, Purdue University Graduate School.

[36] Gonzalez de Gortari, M., *Improving the role of polyphthalamide (PPA) in transport applications by increasing its sustainability and mechanical properties through the use of a sustainable polyamide (PA410) and renewable biocarbon.* 2022, University of Guelph.

[37] Siengchin, S., A review on lightweight materials for defence applications: A present and future developments. *Defence Technology,* 2023.

[38] Diniță, A., et al., Advancements in Fiber-Reinforced Polymer Composites: A Comprehensive Analysis. *Polymers,* 2023. 16(1): p. 2.

[39] Tuli, N. T., S. Khatun, and A. B. Rashid, Unlocking the future of precision manufacturing: A comprehensive exploration of 3D printing with fiber-reinforced composites in aerospace, automotive, medical, and consumer industries. *Heliyon,* 2024.

[40] Balint, T., et al., Underutilized agricultural co-product as a sustainable biofiller for polyamide 6, 6: Effect of carbonization temperature. *Molecules,* 2020. 25(6): p. 1455.

[41] Oulidi, O., et al., Peanut shell from agricultural wastes as a sustainable filler for polyamide biocomposites fabrication. *Journal of King Saud University-Science,* 2022. 34(6): p. 102148.

[42] de Melo, R. P., Development of polyamide composites with natural fibers for automotive applications. 2015, Ecole Nationale Supérieure des Mines de Paris; Universidade federal do Rio ….

[43] Birch, A., et al. Development of cost effective and sustainable polyamide blends for automotive applications. in *SPE Automotive Composites Conference & Exhibition.* 2015.

[44] Bledzki, A. K., O. Faruk, and A. Jaszkiewicz, Cars from renewable materials. *Kompozyty,* 2010. 10(3): p. 282-288.

[45] Andresen, C., et al., *Biobased automobile parts investigation.* A report developed for the USDA Office of energy policy and new uses, 2012.

[46] Muukka, S., *Bio-oil based polymeric composites for additive manufacturing.* 2020.

[47] Shah, V., et al., Advances in polymeric nanocomposites for automotive applications: A review. *Polymers for Advanced Technologies,* 2022. 33(10): p. 3023-3048.

[48] Chowdhury, M. I. S., et al., 11 - Polymer nanocomposites for automotive applications, in *Advanced Polymer Nanocomposites,* M. E. Hoque, K. Ramar, and A. Sharif, Editors. 2022, Woodhead Publishing. p. 267-317.

[49] Oladele, I. O., T. F. Omotosho, and A. A. Adediran, Polymer-based composites: an indispensable material for present and future applications. *International Journal of Polymer Science,* 2020. 2020: p. 1-12.

[50] Karak, N., Fundamentals of nanomaterials and polymer nanocomposites, in *Nanomaterials and polymer nanocomposites*. 2019, Elsevier. p. 1-45.

[51] Valério, M. B., et al., Barrier behavior evaluation of polyamide 12 nanoclay composite to offshore application by electrochemical impedance spectroscopy. *Journal of Applied Polymer Science,* 2022. 139(13): p. 51863.

[52] Grätzl, T., et al., Influence of the automotive paint shop on mechanical properties of continuous fibre-reinforced thermoplastics. *Composite Structures,* 2019. 208: p. 557-565.

[53] Özbay, B. and İ. E. Serhatlı, Processing and characterization of hollow glass-filled polyamide 12 composites by selective laser sintering method. *Materials Technology,* 2022. 37(4): p. 213-223.

[54] Chen, P., et al., Investigation into the processability, recyclability and crystalline structure of selective laser sintered Polyamide 6 in comparison with Polyamide 12. *Polymer Testing,* 2018. 69: p. 366-374.

[55] Pietroluongo, M., et al., Mechanical recycling of an end-of-life automotive composite component. *Sustainable Materials and Technologies,* 2020. 23: p. e00143.

[56] Caltagirone, P. E., et al., Substitution of virgin carbon fiber with low-cost recycled fiber in automotive grade injection molding polyamide 66 for equivalent composite mechanical performance with improved sustainability. *Composites Part B: Engineering,* 2021. 221: p. 109007.

[57] Chanda, M., Chemical aspects of polymer recycling. *Advanced Industrial and Engineering Polymer Research,* 2021. 4(3): p. 133-150.

[58] Dabees, S., T. Osman, and B. M. Kamel, Mechanical, thermal, and flammability properties of polyamide-6 reinforced with a combination of carbon nanotubes and titanium dioxide for under-the-hood applications. *Journal of Thermoplastic Composite Materials,* 2023. 36(4): p. 1545-1575.

[59] Chowdhury, A., Advances in Halogen-Free Flame Retardants. *Latest Trends in Textile and Fashion Designing,* 2018. 1.

[60] Feuchter, H., F. Poutch, and A. Beard, The impact of halogen free phosphorus, inorganic and nitrogen flame retardants on the toxicity and density of smoke from 10 common polymers. *Fire and Materials,* 2023. 47(8): p. 1003-1023.

[61] Zhang, S., et al., An inherently flame-retardant polyamide 6 containing a phosphorus group prepared by transesterification polymerization. *Polymer,* 2020. 207: p. 122890.

[62] Oliveira, F. M., T. A. Ezquerra, and Z. Z. Denchev, Broadband electrical conductivity of metal/carbon nanotubes polyamide 6 composites fabricated by reactive encapsulation. *Journal of Materials Science,* 2024. 59(4): p. 1348-1363.

[63] Luiz, G., G. Oliveira, and L. B. Canto, Development of polyamide 6-graphene oxide nanocomposite and direct-joining with aluminum alloy for lightweight engineering applications. *Polymer Engineering & Science,* 2024. 64(2): p. 663-674.

[64] Armioun, S., et al., Sustainable and lightweight biopolyamide hybrid composites for greener auto parts. *The Canadian Journal of Chemical Engineering,* 2016. 94(11): p. 2052-2060.

[65] Czel, G., A. Sycheva, and D. Janovszky, Effect of different fillers on thermal conductivity, tribological properties of Polyamide 6. *Scientific Reports,* 2023. 13(1): p. 845.

[66] Aydemir, D., et al., Heat treated wood–nylon 6 composites. *Composites Part B: Engineering,* 2015. 68: p. 414-423.

[67] Feldmann, M., H. P. Heim, and J. C. Zarges, Influence of the process parameters on the mechanical properties of engineering biocomposites using a twin-screw extruder. *Composites Part A: Applied Science and Manufacturing,* 2016. 83: p. 113-119.

[68] Rahman, M. Z., et al., Advanced biopolymers for automobile and aviation engineering applications. *Journal of Polymer Research,* 2023. 30(3): p. 106.

[69] Kuciel, S., P. Koźniar, and A. Liber-Kneć, Polyamides from renewable sources as matrices of short fiber reinforced biocomposites. *Polimery,* 2012. 57(9): p. 627-634.

[70] Brehmer, B., Polyamides from biomass derived monomers. *Bio-Based Plastics: Materials and Applications,* 2013: p. 275-293.

[71] Chakraborty, I. and K. Chatterjee, Polymers and composites derived from castor oil as sustainable materials and degradable biomaterials: current status and emerging trends. *Biomacromolecules,* 2020. 21(12): p. 4639-4662.

[72] Fan, J., et al., Castor oil-based multi-functional monomers and their application in polyamide design. *Industrial Crops and Products,* 2023. 203: p. 117188.

[73] Yan, K., et al., Bio-based monomers for amide-containing sustainable polymers. *Chemical Communications,* 2023. 59(4): p. 382-400.

[74] Armioun, S., et al., Biopolyamides and High-Performance Natural Fiber-Reinforced Biocomposites. *Handbook of Composites from Renewable Materials,* 2017: p. 253-270.

[75] Sergi, C., et al., Toughened Bio-Polyamide 11 for Impact-Resistant Intraply Basalt/Flax Hybrid Composites. *Macromol,* 2022. 2(2): p. 154-167.

[76] Wu, T., et al., Identification of two polyamides (PA11 and PA1012) using pyrolysis-GC/MS and MALDI-TOF MS. *Polymer Testing,* 2013. 32(2): p. 426-431.

[77] Sarfraz, M. S., H. Hong, and S. S. Kim, Recent developments in the manufacturing technologies of composite components and their cost-effectiveness in the automotive industry: A review study. *Composite Structures,* 2021. 266: p. 113864.

Chapter 2

Thermomechanical Properties of Nanocomposites

Sabarish Radoor[1,*]
M. Vintu[2]
and Jasila Karayil[3]

[1]Materials and Production Engineering, The Sirindhorn International Thai-German Graduate School of Engineering (TGGS), King Mongkut's University of Technology North Bangkok, Bangkok, Thailand
[2]Department of Chemistry, National Institute of Technology, NIT Campus, Calicut, Kerala, India
[3]Department of Applied Science, Government Engineering College West Hill, Kozhikode, Kerala, India

Abstract

Nanofillers possess unique characteristics, such as high surface area, small pores, and remarkable thermal, electrical, and mechanical properties, making them ideal for modifying the properties of the composites. The properties of the nanocomposite material will be influenced by the nanoparticle type, concentration, and processing techniques. This chapter thoroughly discusses the effects of nanoparticles (zeolite, TiO_2, nanoclay, and nanosilica) on the thermo-mechanical properties of polymer nanocomposite. Surface treatment of nanofillers enhances their interfacial interaction with the polymer matrix, leading to improvements in properties such as tensile strength, elastic modulus, flexural strength, and impact resistance. Therefore, this chapter also

---

* Corresponding Author's Email: sabarishchem@gmail.com

In: Mechanical and Thermal Properties of Engineering …
Editors: Senthilkumar Krishnasamy, Thitinun Ungtrakul, Sabarish Radoor et al.
ISBN: 979-8-89530-460-0

discusses surface-treated nanofiller-incorporated polymer composites and their enhanced thermo-mechanical performance.

Keywords: thermal stability, mechanical properties, nanocomposites, zeolite, TiO_2, nano clay, nano silica

1. Introduction

Polymer nanocomposites open up new dimensions in composite materials, offering superior properties compared to conventional micro/macro composites, particularly in terms of thermo-mechanical performance, oxidative stability, optical and electrical properties, and durability [1, 2]. Inorganic filler-embedded polymer nanocomposites exhibit remarkable features due to their unique characteristics and broad range of applications across various fields [3, 4]. The distinct properties of polymer nanocomposites are primarily determined by the characteristics of the integrated inorganic nanofiller and the polymer matrix used.

Polymeric materials have gained significant attention from scientists due to their exceptional photo-thermal resistance, environmental stability, and enhanced mechanical, electrical, and chemical properties. These attributes make them highly desirable for a wide range of applications, driving substantial research interest in this field [1, 5]. However, some polymers, such as epoxy resins, are inherently brittle, which limits their use in applications requiring high fracture and impact strength. The incorporation of nanofillers into polymer matrices has led to dramatic improvements in the mechanical performance of polymer nanocomposites. The large surface area of nanofillers enables strong interactions with the polymer matrix, enhancing the material's overall properties [6]. Reinforced polymer composites are also cost-effective, and global efforts are currently underway to develop and characterize composite systems that address existing limitations and broaden their applications [7, 8]. The properties of polymer composites are influenced by the type of polymer and the proportion of filler used. By selecting suitable nanofillers, the mechanical (tensile strength, tensile modulus), flexural, impact, fracture, thermo-electrical, and magnetic properties of the polymer can be significantly modified [9, 10]. Figure 1 illustrates a schematic diagram depicting the incorporation of nanoparticles into a polymer matrix. There are various synthetic methods to produce polymeric nanocomposites, including in situ polymerization [11], mechanical mixing, modified emulsion polymerization, sol-gel processes, solution dispersion, melt compounding, melt extrusion, injection molding, and selective laser sintering. While each method has its specific approach, the final outcome and structural morphology of the nanocomposites largely dictate their potential applications. The morphological features of nanocomposites depend not only on the preparation method but also on the synergistic interactions between the polymer and nanoparticles. These interactions ensure the uniform distribution of nanoparticles within the polymer matrix, which is critical for achieving desired propertie [12, 13].

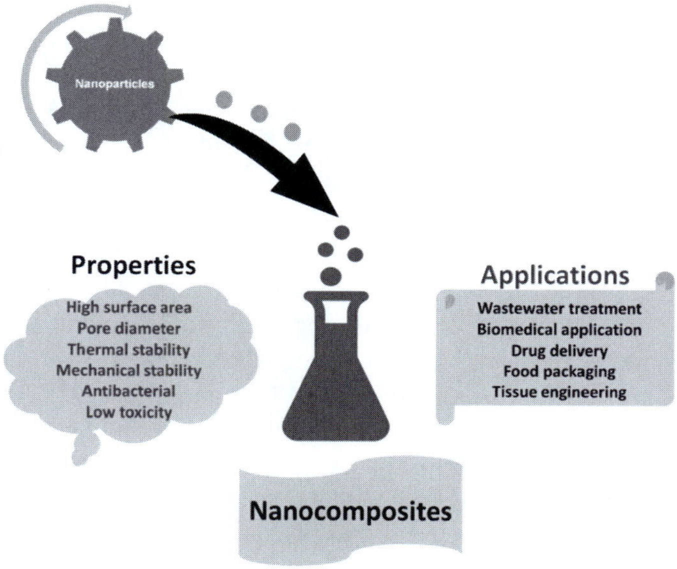

Figure 1. Schematic illustration of nanoparticles incorporated polymer matrix with properties and applications.

2. Thermo-Mechanical Properties of Zeolite-Based Polymer Composites

Zeolites are an aluminosilicate mineral with three-dimensional networks of microscopic pores [14, 15]. The thermal, mechanical, and chemical stability of the composites could be increased by the incorporation of zeolite [16, 17]. Consequently, there is increasing interest in developing polymer composites that utilize zeolites for various applications [18, 19]. Chitosan has a limited mechanical stability; however, its properties could be increased with the addition of filler. Ghosh and a colleague [20] used zeolite molecular sieve 5A as a reinforcing filler to improve the thermomechanical properties of chitosan, which typically exhibits limited mechanical stability. They varied the zeolite weight percentage from 2 to 10 wt%, and it was observed that the tensile strength and modulus increased with increasing zeolite concentration. The chitosan composite containing 10 wt% zeolite showed a remarkable improvement in its mechanical properties. Specifically, the tensile strength increased by 2.4 times, while the modulus increases to 21-fold. These improvements are largely due to the strong interaction between the zeolite particles and the chitosan matrix. On the other hand, the addition of zeolite significantly enhances the thermal stability of chitosan composites. Zeolite not only shifts the decomposition temperatures to higher ranges but also reduces the overall mass loss during thermal decomposition. The strong interactions between zeolite and chitosan, combined with the heat-resistant Si–O framework of zeolite, contribute to the formation of protective barriers, restricting polymer motion and improving thermal stability. These findings suggest that chitosan/zeolite composites have great potential for applications requiring enhanced thermomechanical properties. Metin and their group examined the impact of zeolite loading and surface modification using polyethylene glycol and silane coupling agents on the mechanical and structural properties of polypropylene-zeolite composites [21]. The addition of zeolite raised the composite's Young's modulus because the presence of zeolite restricted the flexibility of the polymer chain. Additionally, silane treatment increases the Young's modulus to its maximum. The maximum Young's modulus was observed at a silane concentration of 1 wt%; however, as the silane concentration was increased above that point, the surface modifier's plasticizing impact caused the Young's modulus to decrease. For 3-aminopropyltriethoxysilane (AMPTES), methyltriethoxysilane (MTES) and 3-mercaptopropyltrimethoxysilane (MPTMS), the tensile stress at break reaches its highest value at 1 and 0.5 weight percent coupling agent concentrations, respectively. However, the inclusion of zeolite causes a decrease in tensile strength due to the reduction in the matrix cross-section as well as the creation of voids.

According to Matykiewicz et al., [22] incorporating silsesquioxanes and zeolite enhances the thermomechanical properties of epoxy-basalt fiber composites. While the tensile strength remained relatively unchanged, with the neat basalt epoxy fiber at 300 MPa and little difference after adding zeolite. This finding is supported by morphological analysis. The analysis reveals no fiber pullout, indicating strong interfacial contact between the fibers and the polymer matrix. The images clearly show that zeolites are evenly distributed throughout the matrix. Moreover, the addition of zeolite and silsesquioxanes improved the surface roughness of the fibers, thereby strengthening the bond between the fibers and the matrix, which enhances the composite's impact resistance. Additionally, these additions also contribute to improved thermal stability of the composite. Meanwhile, zeolites have been successfully used to improve the thermomechanical properties of wood-polymer composites (WPCs). The hydrophilic nature of wood flour makes it highly prone to absorbing water molecules, which leads to weak interfacial adhesion between the wood flour and the polymer matrix. The results indicate that the thermo-mechanical and physical properties of wood-polymer composites (WPCs) are influenced by zeolite and pine wood flour. It was shown that zeolite addition decreased the ability of WPCs to adsorb water. However, with an increase in zeolite content from 10 to 50%, WPCs' flexural strength and modulus of elasticity values drop. This phenomenon occurs because zeolite agglomeration, can significantly diminish the mechanical properties of the material. Note that higher amounts of zeolite had higher residue at 500°C because of the presence of silicate on the surface of zeolite, which increased the thermal stability of WPC with increasing zeolite percentage [23].

3. Thermo-Mechanical Properties of Nano Clay-Based Polymer Composites

Biomaterials are increasingly being used as fillers in polymer composites. Clay is one of the most prevalent materials on the earth's surface, having appealing qualities such as high surface area, pore size, strong thermomechanical properties, affordability, non-toxicity, and accessibility. Clay-based composites have drawn a lot of interest in the scientific community [24, 25]. In their study, Kumari et al. [26] compared the effects of pure and surfactant-modified clays (bentonite, vermiculite, and montmorillonite) regarding the mechanical and thermal properties of PVC (polyvinyl chloride). The presence of clay significantly improved the thermal stability of the samples, leading to an increase in PVC's breakdown temperature. Additionally, the mechanical properties, including tensile strength and Young's modulus, were enhanced, demonstrating greater resistance to tensile stress. This improvement is attributed to the high compatibility between the clay and PVC polymers, along with the uniform dispersion of clay within the polymeric matrix. The mechanical characteristics of clay-modified PVC composite films benefited from strong interfacial adhesion between the clay and the polymer. Solution casting was employed to successfully produce a chitosan-based nanocomposite with varying concentrations of clay, both with and without glycerol. Due to the strong interactions between chitosan and clay, the clay-incorporated chitosan nanocomposites exhibit significantly enhanced thermal stability compared to pure chitosan films. During the pyrolysis process, the nanolayer silicates from the clay act as a barrier, effectively preventing the degradation of volatile compounds and limiting the breakdown of chitosan. The maximum tensile strength was observed at a clay content of 5 wt%; however, as the clay concentration increased beyond this point, the tensile strength began to decrease due to the tendency of clay to agglomerate within the chitosan matrix. (Figure 2). The homogenous dispersion of clay particles throughout the matrix leads to strong interaction with the polymer, additionally clay can interact through electrostatic interactions with the chitosan $-NH_2$ and $-NH_3$ groups. Tensile modulus showed a similar pattern, with the highest value being attained at 5wt%, of clay. The nanocomposite films made of chitosan and clay also possess good antibacterial properties [27].

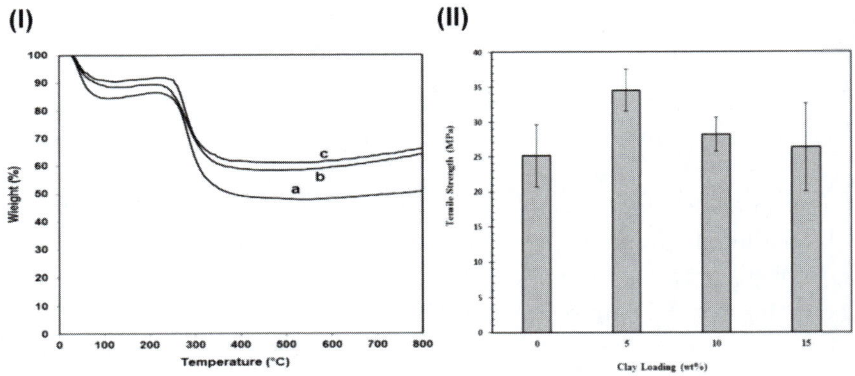

Figure 2. (I) TGA curve of neat chitosan (a), 5wt% loaded clay on chitosan (b), 5 wt% of clay and 20 wt% of glycerol loaded on chitosan (II) tensile strength of different percentage of clay loaded chitosan nanocomposite.

Sabu and co-workers [28] examined the hydrophilic, thermomechanical behavior of the material of hybrid epoxy composites filled with organoclay. Significant thermal stability is demonstrated by polyurethane epoxy loaded with 1, 3, and 5 wt% OMMT clay, the degradation temperatures at T60 are as follows: 485, 517, and 551, respectively. Furthermore, the incorporation of clay prevents the volatile components of epoxy clay nanocomposites from degrading and thereby increase thermal stability. The tensile strength and tensile modulus of epoxy nanocomposites increases to 3.5, 7.7 and 22% and 5.7, 4.5 and 23%, respectively.

4. Thermo-Mechanical Properties of Silica-Based Polymer Composites

Silica nanoparticles composed of silicon dioxide have been commonly used in material applications. Due to their special qualities including chemical stability, thermal stability, high surface area, optical characteristics, mechanical, electrical, toughness, etc., they have a wide range of applications in many fields, such as wastewater, medicine, electronics, *etc*. Payraz et al. [29] reported the influence of PVA and silica on the thermomechanical characteristics of nano fibrillated cellulose composites treated with Celluclast. Using casting methods, three distinct nanocomposite films nanofibrillated cellulose (CNF), nanocellulose-PVA (CNF-P), and nanocellulose-silica (CNF-Si) were developed. Degradation of all samples occurred between 250°C and 400°C, which is attributed to the degradation of the cellulose polymer. CNF and CNF-Si showed single degradation within this range, while CNF-P continued to degrade up to 500°C, likely due to interactions between PVA and cellulose fibers, as supported by peaks at 2916 cm^{-1} and 835 cm^{-1} in FT-IR analysis. CNF-Si had the lowest thermal stability, with its degradation occurring at lower temperatures due to silica's higher surface area and amorphous structure, which accelerated heat transfer and reduced stability. The CNF-P film displayed superior tensile strength and Young's modulus compared to the CNF-Si film due to hydrogen bonding between PVA and cellulose fibers. This bonding reduced chain segment mobility, enabling better load transfer and enhancing tensile strength. In contrast, CNF-Si exhibited lower mechanical performance Khankrua et al. [30] examined the thermomechanical properties of polyester/silica nanocomposites. The thermal analysis indicates that the inclusion of silica nanoparticles raised both onset and inflection temperatures of all three samples (PLA, PHBV, and PBS). This is attributed to the high surface area of the silica nanoparticles, which increases the interfacial interaction with the polymer matrix, allowing thermal energy to transfer from the polymer to the filler particles. The elongation at break (%) of polylactic acid (PLA) shows a slight increase with low silica content (0.1-

0.5 wt%), a trend also observed in polybutylene succinate (PBS) and polyhydroxybutyrate-valerate (PHBV) samples. This improvement is attributed to well-dispersed nanoparticles, which enhance the material's strength. However, when silica content exceeds 0.5 wt%, a decline in elongation at break (%) is noted across all polymer types. This reduction is attributed to the formation of silica nanoparticle agglomerates, which act as stress concentration sites rather than contributing to the toughening of the polymer matrix, as evidenced by SEM micrographs. Consequently, the mechanical performance of the polymer nanocomposites diminishes with higher silica content. Meanwhile, the effect of weight percentages of nano silica (0.3 wt%, 0.75 wt%, and 1 wt%) on the mechanical properties of nano silica-vinyl ester nanocomposites were studied by Baniani et al. [31]. It has been found that adding silica fillers increases the Young's modulus while a decrease in tensile strength was noted. This is attributed to the fact that high silica particles are added in excess, the strength will decrease as a result of the weak interfacial connection.

5. Thermo-Mechanical Properties of Titanium Dioxide-Based Polymer Composites

The interaction between nanofillers and polymers enables effective confinement of the polymer at the polymer-nanofiller interface. This interaction is a key factor in enhancing the mechano-chemical properties of these polymers, making them superior to traditional polymers. Among the diverse range of fillers used in various industries, titanium dioxide (TiO_2) holds a significant position in polymer composites, setting itself apart from other inorganic fillers. TiO_2-based polymer composites are widely utilized for synthesizing malleable, high-performance polymer structures suitable for 3D printing. The improved photocatalytic properties of TiO_2 allow for its application in various fields, such as the degradation of organic contaminants and pathogens, the development of antiseptic and antibacterial formulations, and as a durable UV-resistant material. This versatility stems from its cost-effectiveness, non-toxic characteristics, chemical inertness, and high refractive index. Additionally, TiO_2 serves as an essential component in pharmaceuticals, cosmetics, paints, drug delivery systems, solar cells, chemical sensors, luminescent materials, and photocatalysts. [32-35].

Polymeric matrices infused with titanium dioxide (TiO_2) exhibit improved physical and mechanical properties. Compared to the pristine base matrix,

TiO_2-filled polymeric nanocomposites show enhanced mechanical strength and modulus. The internal structure of these TiO_2 nanocomposites plays a crucial role in fine-tuning their mechanical properties. However, particle aggregation and/or agglomeration can lead to a reduction in mechanical performance due to the defects created within the nanocomposites. Recently, different types of one-dimensional nanomaterials (nanorods, nanotubes, nanoribbons) with a high aspect ratio have gained much attention for the synthesis of uniformly dispersed polymer composites [36]. There are many factors that may affect the mechanical properties of composites, those are presented in Figure 3.

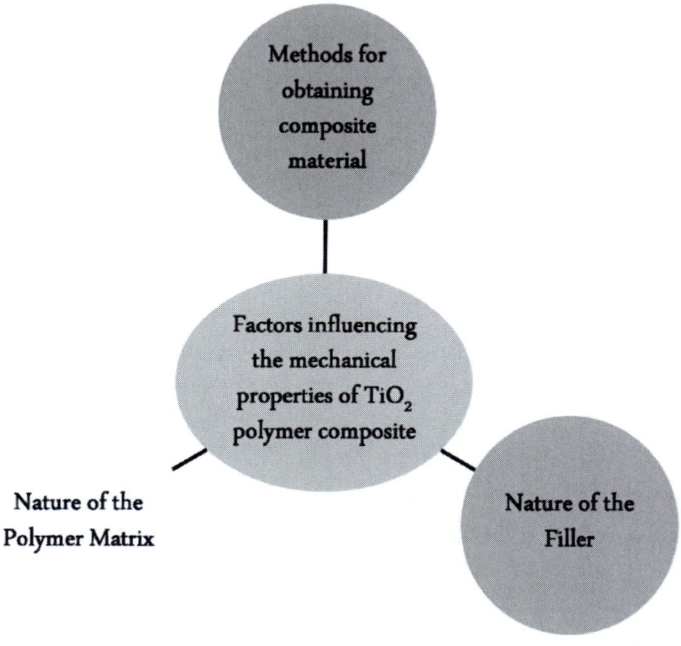

Figure 3. Factors affecting the mechanical properties of composites.

The mechano-chemical properties of polymers can be significantly influenced by various factors related TiO_2, such as the amount used, its size, and shape. The composite strength can be enhanced either by reducing the particle size or by increasing the particle volume fraction. Additionally, good dispersibility of TiO_2 within the polymer matrix improves thermal stability. Awang et al. [37] made a thorough study on polypropylene-rice husk green composites. The alteration of mechanical and thermal properties was evaluated by the addition of TiO_2 and the newly prepared composite exhibited better mechanical properties than the composites without TiO_2 and pure PP. The formulation PP/RH-10wt%/TiO_2 achieved a maximum tensile strength of 41.2 MPa.

Sreekumar et al. [38] conducted a study on 1:1 starch/polyvinyl alcohol (PVA) composites, fine-tuning their morphological and thermo-mechanical properties through the incorporation of nano-titanium dioxide (TiO_2). The results indicated that at higher concentrations of nano-TiO_2, tensile strength improved, while the percentage elongation and energy at break gradually decreased. Bora et al. [39] investigated polyphenylenesulphide (PPS)-TiO_2 composites, examining how the flexural and solid particle erosive wear properties varied with the weight percentage of TiO_2 particles (up to 25 wt.%). As the distribution of TiO_2 particles in the PPS matrix increased, the stiffness of the composite also improved. The maximum values for flexural and residual flexural strength were observed at 10 wt.% TiO_2 concentration Saluja et al., demonstrated improvements in thermal conductivity and glass transition temperature (Tg) in polyester-TiO_2 composites by incorporating various concentrations of filler particles into the polyester matrix. In most cases, the mechanical properties of the composites, including flexibility, ductility, hardness, strength, and stiffness, were enhanced through the doping of nanofillers.

A wide variety of polymer matrices have been used for the preparation of polymer–TiO_2 nanocomposites, such as polystyrene, polyimide, polyacrylate, polyolefins, poly (methyl methacrylate). Thermosetting polymers i.e.; unsaturated polyester, polyamide, polycarbonate also play a vital role. To enhance the performance of polymer matrices, it is important to maintain a relatively low weight percentage of filler (between 0.5% and 5%), as this can significantly affect interfacial interactions. When optimizing the properties of new polymer-TiO_2 nanoparticle composites, several factors need to be considered, including the characteristics of both the filler and the polymer matrix, the amount of filler incorporated, the distribution of the filler (ensuring no agglomeration), the preparation method, and the concentration of the modifying coupling agent at the filler-matrix interface. Metanawin et al. [40] investigated the impact of TiO_2 on poly(butylene terephthalate) (PBT) composite fibers, evaluating their mechano-thermal properties. Their results revealed that an increased quantity of TiO_2 resulted in a reduction in the tensile strength of the TiO_2/PBT fibers within the polymer matrix.

In recent years, the integration of TiO_2 nanomaterials into polymer matrices to form nanocomposites has become an exciting approach for a wide range of applications. This integration can significantly affect the glass transition temperature (Tg) of both polymer matrices and polymer nanocomposites [41]. The thermal properties of these nanocomposites are not only of practical importance but also play a crucial role in determining operational temperatures, mechano-physical properties, and thermal stability. The diverse nature of polymer material and nanofillers make the composites in a premium position. Their generalized form can be classified as a heat-conducting filler in a non-heat-conductive matrix and non-conductive filler in a non-conductive matrix. Changes in temperature transitions have been noticed recently in nanocomposites and bare polymer matrices due to this form.

Thermogravimetric analysis (TGA) was used to evaluate thermal stability and the interactions among polymer matrix particles. Generally, TiO_2-filled polymer composites demonstrate higher degradation temperatures. The enhanced adhesion at the polymer matrix/TiO_2 interface is a critical factor contributing to this increase in degradation temperature. Strong interfacial interactions restrict the movement of polymer chains, making chain scission more challenging at lower temperatures. As a result, the degradation temperature of the nanocomposite shifts to a higher range [42]. Esthappan et al. [43]. investigated the thermo-mechanical properties of polypropylene/titanium dioxide nanocomposite fibers and found that the interaction between nanoparticles and the polymer backbone enhances the properties of polypropylene. They also examined the interfacial interactions between bamboo fibers (BFs) and polypropylene (PP) by modifying BFs with TiO_2 nanoparticles. Their results indicated that incorporating 0.4 wt% TiO_2 nanoparticles improved the static and dynamic mechanical properties, thermal stability, and water resistance of the composites [41]. The crystallization behavior of polymers and their composites can be analyzed using Differential Scanning Calorimetry (DSC). The influence of TiO_2 nanoparticles on the crystallization characteristics of polymers is assessed through DSC experiments. Differences in crystallization temperatures (Tc), melting temperatures (Tm), and the corresponding enthalpies of the composites compared to their parent polymers are attributed to changes mediated by the transition from bulk to the interface.

Conclusion

In this chapter, we discussed the effect of various filler materials—specifically TiO_2, zeolite, silica, and nano clay on the thermal and mechanical characteristics of composites. Research indicates that the incorporation of nanofillers can lead to significant enhancements in the thermomechanical properties of composite materials. This improvement is largely due to two key factors: the excellent dispersion of the fillers within the polymer matrix and the strong interfacial interactions that develop between the fillers and the polymer. Furthermore, the chemical modification of fillers plays a critical role in optimizing these properties. By altering the surface chemistry of the fillers, it is possible to enhance their compatibility with the polymer matrix, which can lead to even greater improvements in performance. Moreover, the presence of nanofillers has been shown to enhance the thermal conductivity of composite materials, making them more effective in applications that require efficient heat dissipation.

References

[1] Mittal, V., *Optimization of Polymer Nanocomposite Properties,* 2010.

[2] Radoor, S., J. Karayil, J. M. Shivanna, A. Jayakumar, J. Parameswaranpillai, S. Siengchin, Cotton fibers, their composites and applications, in: P*lant Fibers, their Composites, and Applications,* 2022, pp. 379-390.

[3] Schadler, L. S., L. C. Brinson, W. G. Sawyer, Polymer nanocomposites: A small part of the story, *Jom,* 59 (2007) 53-60.

[4] Radoor, S., J. Karayil, A. Jayakumar, J. Parameswaranpillai, S. Siengchin, PLA based bionanocomposites for food packaging applications, in: *Bionanocomposites for Food Packaging Applications,* 2022, pp. 115-133.

[5] Radoor, S., J. Karayil, A. Bemplassery, A. Jayakumar, J. Parameswaranpillai, S. Siengchin, Spectroscopy of elastomer blends and composites, in: *Elastomer Blends and Composites,* 2022, pp. 195-207.

[6] Young, R. J., I. A. Kinloch, L. Gong, K. S. Novoselov, The mechanics of graphene nanocomposites: *A review, Composites Science and Technology*, 72 (2012) 1459-1476.

[7] Cazan, C., A. Enesca, L. Andronic, Synergic Effect of TiO_2 Filler on the Mechanical Properties of Polymer Nanocomposites, *Polymers,* 13 (2021).

[8] Radoor, S., J. Karayil, A. Jayakumar, E. Krishnankutty Radhakrishnan, J. Parameswaranpillai, S. Siengchin, Alginate-based bionanocomposites in wound dressings, in: *Bionanocomposites in Tissue Engineering and Regenerative Medicine,* 2021, pp. 351-375.

[9] Rubab, Z., A. Afzal, H. M. Siddiqi, S. Saeed, Preparation, Characterization, and Enhanced Thermal and Mechanical Properties of Epoxy-Titania Composites, *The Scientific World Journal,* 2014 (2014) 1-7.

[10] Radoor, S., J. M. Shivanna, S. Baswaraj, C. R. Chowdhury, A. Jayakumar, J. Karayil, J. T. Kim, J. Lee, S. Siengchin, Lightweight and sustainable materials for food packaging applications, in: *Lightweight and Sustainable Composite Materials,* 2023, pp. 219-240.

[11] Radoor, S., J. Karayil, S. Damodaran, A. Jayakumar, J. Parameswaranpillai, S. Siengchin, Bio Nanocomposite Films in the Food Packaging Applications, in: *Polymer Based Bio-nanocomposites,* 2022, pp. 255-273.

[12] Radoor, S., J. Karayil, R. Soman, A. Jayakumar, E. K. Radhakrishnan, J. Parameswaranpillai, S. Siengchin, Influence of Nanoclay on the Thermal Properties of the Natural Fiber-Based Hybrid Composites, in: *Natural Fiber-Reinforced Composites,* 2022, pp. 239-254.

[13] Radoor, S., A. Bemplassery, A. Jayakumar, J. Karayil, J. M. Shivanna, J. T. Kim, J. Lee, J. Parameswaranpillai, S. Siengchin, Natural Fiber/Epoxy-Based Hybrid Composites, in: *Epoxy-Based Biocomposites*, 2023, pp. 177-190.

[14] Sabarish, R., G. Unnikrishnan, Synthesis, characterization and catalytic activity of hierarchical ZSM-5 templated by carboxymethyl cellulose, *Powder Technology,* 320 (2017) 412-419.

[15] Sabarish, R., G. Unnikrishnan, Synthesis, characterization and evaluations of micro/mesoporous ZSM-5 zeolite using starch as bio template, *SN Applied Sciences,* 1 (2019).

[16] Radoor, S., J. Karayil, A. Jayakumar, D. Nandi, J. Parameswaranpillai, J. Lee, J. M. Shivanna, R. Nithya, S. Siengchin, Adsorption of Cationic Dye onto ZSM-5 Zeolite-Based Bio Membrane: Characterizations, Kinetics and Adsorption Isotherm, *Journal of Polymers and the Environment,* 30 (2022) 3279-3292.

[17] Radoor, S., J. Karayil, A. Jayakumar, D. Nandi, J. Parameswaranpillai, J. Lee, S. Siengchin, Polyvinyl alcohol/guar gum-based bio-adsorbent for the removal of cationic and anionic dyes from aqueous solution, *Polymer Bulletin,* 80 (2022) 10165-10191.

[18] Sabarish, R., K. Jasila, J. Aswathy, P. Jyotishkumar, S. Suchart, Fabrication of PVA/agar/modified ZSM-5 zeolite membrane for removal of anionic dye from aqueous solution, *International Journal of Environmental Science and Technology,* 18 (2020) 2571-2586.

[19] Radoor, S., J. Karayil, A. Jayakumar, J. Lee, D. Nandi, J. Parameswaranpillai, B. Pant, S. Siengchin, Efficient Removal of Organic Dye from Aqueous Solution Using Hierarchical Zeolite-Based Biomembrane: Isotherm, Kinetics, Thermodynamics and Recycling Studies, *Catalysts,* 12 (2022).

[20] Ghosh, A., L. Ma, C. Gao, Zeolite molecular sieve 5A acts as a reinforcing filler, altering the morphological, mechanical, and thermal properties of chitosan, *Journal of Materials Science,* 48 (2013) 3926-3935.

[21] Metın, D., F. Tihminlioğlu, D. Balköse, S. Ülkü, The effect of interfacial interactions on the mechanical properties of polypropylene/natural zeolite composites, *Composites Part A: Applied Science and Manufacturing,* 35 (2004) 23-32.

[22] Matykiewicz, D., K. Lewandowski, B. Dudziec, Evaluation of thermomechanical properties of epoxy–basalt fibre composites modified with zeolite and silsesquioxane, *Composite Interfaces,* 24 (2016) 489-498.

[23] Kaymakci, A., T. Gulec, S. K. Hosseinihashemi, N. Ayrilmis, Physical, mechanical and thermal properties of wood/zeolite/plastic hybrid composites, Maderas. *Ciencia y tecnología,* (2017) 0-0.

[24] Yadav, V. B., R. Gadi, S. Kalra, Clay based nanocomposites for removal of heavy metals from water: A review, *Journal of Environmental Management,* 232 (2019) 803-817.

[25] Ruiz-Hitzky, E., M. Darder, A. C. S. Alcântara, B. Wicklein, P. Aranda, Recent Advances on Fibrous Clay-Based Nanocomposites, in: *Organic-Inorganic Hybrid Nanomaterials,* 2014, pp. 39-86.

[26] Kumari, N., C. Mohan, A. Negi, An Investigative Study on the Structural, Thermal and Mechanical Properties of Clay-Based PVC Polymer Composite Films, *Polymers,* 15 (2023).

[27] Kusmono, I. Abdurrahim, Water sorption, antimicrobial activity, and thermal and mechanical properties of chitosan/clay/glycerol nanocomposite films, *Heliyon,* 5 (2019).

[28] Sabu, M., Y. J. V. Ruban, M. S. Ginil, Effect of organoclay on contact angle, thermal and mechanical properties of filled epoxy composites, *Materials Today: Proceedings,* 41 (2021) 549-556.

[29] Poyraz, B., A. Tozluoğlu, Z. Candan, A. Demir, M. Yavuz, Influence of PVA and silica on chemical, thermo-mechanical and electrical properties of Celluclast-treated nanofibrillated cellulose composites, *International Journal of Biological Macromolecules,* 104 (2017) 384-392.

[30] Khankrua, R., S. Pivsa-Art, H. Hiroyuki, S. Suttiruengwong, *Thermal and Mechanical Properties of Biodegradable Polyester/Silica Nanocomposites, Energy Procedia,* 34 (2013) 705-713.

[31] Baniani, D. D., S. A. J. Jahromi, S. M. Zebarjad, A study on role of nanosized SiO_2 on deformation mechanism of vinyl ester, *Bulletin of Materials Science,* 37 (2014) 1677-1683.

[32] Vidakis, N., M. Petousis, A. Maniadi, E. Koudoumas, M. Liebscher, L. Tzounis, Mechanical Properties of 3D-Printed Acrylonitrile–Butadiene–Styrene TiO_2 and ATO Nanocomposites, *Polymers,* 12 (2020).

[33] Andronic, L., A. Enesca, Black TiO_2 Synthesis by Chemical Reduction Methods for Photocatalysis Applications, *Frontiers in Chemistry,* 8 (2020).

[34] Uddin, M. J., D. Mondal, C. A. Morris, T. Lopez, U. Diebold, R. D. Gonzalez, An in vitro controlled release study of valproic acid encapsulated in a titania ceramic matrix, *Applied Surface Science*, 257 (2011) 7920-7927.

[35] Fujihara, K., A. Kumar, R. Jose, S. Ramakrishna, S. Uchida, Spray deposition of electrospun TiO2nanorods for dye-sensitized solar cell, *Nanotechnology,* 18 (2007).

[36] DeArmitt, C., R. Rothon, Fillers and surface treatment, *Plastics, Additives and Compounding,* 4 (2002) 12-14.

[37] Awang, M., W. R. Wan Mohd, N. Sarifuddin, Study the effects of an addition of titanium dioxide (TiO2) on the mechanical and thermal properties of polypropylene-rice husk green composites, *Materials Research Express,* 6 (2019).

[38] Sreekumar, P. A., M. A. Al-Harthi, S. K. De, Reinforcement of starch/polyvinyl alcohol blend using nano-titanium dioxide, *Journal of Composite Materials,* 46 (2012) 3181-3187.

[39] Bora, M. Ö., O. Çoban, E. Avcu, S. Fidan, T. Sınmazçelik, The effect of TIO2filler content on the mechanical, thermal, and tribological properties of TiO2/PPS composites, *Polymer Composites,* 34 (2013) 1591-1599.

[40] Zhou, J., T. Metanawin, A. Jamjumrus, S. Metanawin, O. Adiguzel, *Morphology, Mechanical and Thermal Properties of PBT-TiO2Polymer Nanocomposite, MATEC Web of Conferences,* 30 (2015).

[41] Guo, J., M. Cao, W. Ren, H. Wang, Y. Yu, Mechanical, dynamic mechanical and thermal properties of TiO_2 nanoparticles treatment bamboo fiber-reinforced polypropylene composites, *Journal of Materials Science,* 56 (2021) 12643-12659.

[42] Yang, F., G. L. Nelson, Polymer/silica nanocomposites prepared via extrusion, *Polymers for Advanced Technologies,* 17 (2006) 320-326.

[43] Esthappan, S. K., S. K. Kuttappan, R. Joseph, Thermal and mechanical properties of polypropylene/titanium dioxide nanocomposite fibers, *Materials & Design,* 37 (2012) 537-542.

Chapter 3

The Influence of Fibre Surface Treatments on the Mechanical and Thermal Properties of Natural Fibre Reinforced PA6 Composites

Chirasmayee Mohanty
Debasis Nanda
Nigamananda Das
and Ajaya Kumar Behera*
Department of Chemistry, Utkal University, Bhubaneswar, India

Abstract

Composite materials top the list of high-tech industries due to their unique mechanical properties combined with their regular lightweight design, which boosts the use of the materials in the automotive, aerospace, packing, and other businesses. As environmental issues are becoming more and more acute, there is an identified necessity for the transition to more sustainable options, where natural fibre composites are a viable choice. However, despite the renewability frame of fibre materials, certain complications exist due to the hydrophilic properties of natural fibres, which creates obstacles for efficient interaction between it and polymer matrix. This is why surface treatments are indispensable to improve interfacial adhesion and overall performance. The study aimed to research the impact of fibre surface treatment with subsequent thermal characteristics of polyamide composites and its kinematic and strengthening properties. The adhesion between fibre and matrix can be

* Corresponding Author's Email: ajayabehera@utkaluniversity.ac.in, Fax. +91 0674 2581850, ORCID ID: 0000-0002-6172-0867.

In: Mechanical and Thermal Properties of Engineering …
Editors: Senthilkumar Krishnasamy, Thitinun Ungtrakul, Sabarish Radoor et al.
ISBN: 979-8-89530-460-0

improved and physical-mechanical properties strengthened through a combination of various methods that are adequate for alkaline and silane treatment. This may result in the development of composite materials that are environmentally friendly and simultaneously high performing.

Keywords: natural fibre, biodegradation, physical properties, PA6

1. Introduction

In years there has been a focus, on preserving the environment globally driven by a growing awareness of issues like air and water pollution, resource depletion and climate change. There has been a change, in mindset that has led to efforts for advancement by implementing regulations and developing cutting edge technologies to mitigate the impact of activities, on the environment. Friendly materials represent this shift by being designed to reduce the effects of various environmental processes and promote sustainable practices [1-3]. Components based materials, such, as panels are highly valued for their durability and lightweight properties. These materials have found application, across industries including sports, automotive, aerospace and marine sectors over an extended period. In recent years, single-use materials made from fossil fuels have gained considerable attention due to their non-degradable nature, requiring the development of sustainable alternatives. As a result, an interest in researching eco-friendly materials for panels is increasing, in particular composites reinforced with natural fibres [4-6].

Natural fibre reinforced composites mix promising mechanical properties with the sustainability of renewable sources to become an attractive alternative. With materials like jute, hemp, flax, ramie, and sisal becoming the main natural reinforcement candidates, the competition with the amounts of desirable properties including but not confined to high price, biodegradability, supply is more varied. However, many obstacles are still to be conquered for the application of natural composites, and one of the essential considerations is the low compatibility of natural fibres with the vast majority of polymer matrices [7-8]. The hydrophilic property of natural fibres is a key barrier to their efficient use as hydrophobic polymer matrices such as nylon 6, 6 PA6 and nylon 6. The mechanical and thermal properties of these composites are often challenged due to the poor interfacial adhesion of the composites, the dimensional isotropy of the fibres, which causes moisture absorption. It resulted in the necessity of creating surface treatments since it is related to the way to increase composites' performance by achieving an improved matrix-fibre compatibility [9-10].

The research aimed to discover in what way the PA6 natural fibre composites' mechanical and thermal properties were influenced by the fibre surface treatment. It is clear that it enhances the fibre matrix bonding mechanism and diminishes the components in reaction with the absorption moisture process. Moreover, the current research sought to validate the effects of fibre surface treatment on their composites in multiple climatic conditions. Lastly, the study targeted to use detailed analysis and experimentation to answer the gaps in information; thus, answer the broad question how fibre surface treatment and hygrothermal conditions can be used to create advanced composites in examining the specific question [11-12].

2. Natural Fibres

The range of natural fibres is diverse and covers many materials originating from plant, animal, and mineral sources, providing a rich variety of environmentally friendly opportunities.

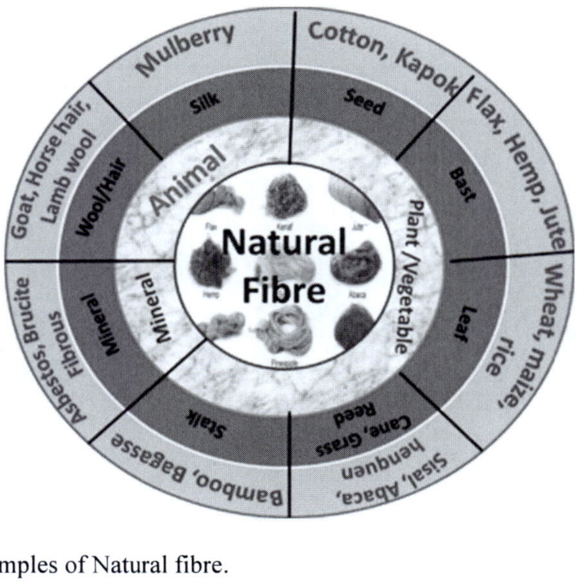

Figure 1. Examples of Natural fibre.

The most extensive list of natural products consists of several relatively common types such as flax, hemp, jute and straw and may appear in a less conventional form such as wood, rice husk and other agricultural wastes, which include wheat and barley, oats, rye etc.

Moreover, fibres from cane, sugar and bamboo, grasses, reeds, kenaf, ramie, oil palm empty fruit bunch, sisal, Raphia, coir, water hyacinth, pennywort, kapok, paper mulberry, banana, pineapple leaf, and papyrus have also been investigated recently [13]. The example of natural fibre is depicted in Figure 1.

2.1. Sources of Natural Fibre

Natural fibres are further categorized as leaf, bast, seed, and fruit fibres depending on their source. The natural fibres have varying properties such as strength, stiffness, roughness and cell length from each other. Cellulose, hemicellulose, lignin, pectin are mostly present in such fibres. The chemical constituents differ based on the type of natural fibre [6]. The schematic classification of natural fibres is shown in Figure 2. These fibres can be broadly divided into three categories: plant, animal, and mineral fibres.

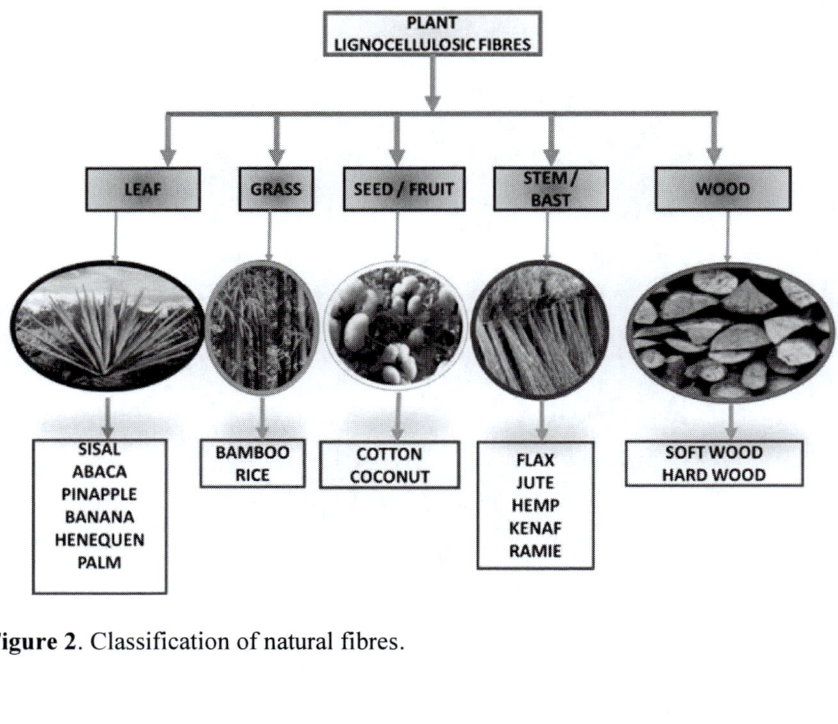

Figure 2. Classification of natural fibres.

2.1.1. Plant Fibres

Plant fibres are mainly made up of lignin, hemicellulose, and cellulose. These fibres are categorized according to where they come from: Fiber from seeds or seed cases (e.g., cotton and kapok); fibre from leaves (e.g., sisal and agave); and fibre from the skin or bast around the stem of the plant (also known as bast or skin fibre). Some of the most highly cherished materials for their incredible tensile strength are employed in robust yarns, textiles, paper, and packaging. Fruit and stalk as well as tree timber provide additional fibre sources. Hemp, cotton, and flax are among the most prevalent natural fibres. Sisal, jute, kenaf, and coconut are all used in a variety of ways [3, 8].

2.1.2. Animal Fibres

Animal fibres are primarily protein-containing materials and are a wide class of natural materials that are often distinguished from synthetic fibres by their specific properties and applications. Silk, wool, angora, mohair, and alpaca are the finest fibres produced by animals. One of the most numerous groups is animal hair, which includes fibres, or rather wool, from several mammals, mostly used for textile purposes. These can be sheep, goats, alpaca and related animals, horse and others. Woolen gastropods are warm and soft materials with good strength, and among others, they remain of the highest quality for the production of clothing and accessories for designers. In addition, using silk fibres, which are obtained from the dried saliva of various insects during the preparation of a cocoon, use of silk is also very popular due to its unique shine and smoothness [12]. Lastly, avian fibres, which are sourced from birds like feathers, extend the variety of animal fibres as an attractive insulating and decorative supplement. Combined, all of these fibres demonstrate the artistry of nature, which has allowed for virtually an infinite variety of possibilities, from haute couture to industrial fabrics [5].

2.1.3. Mineral Fibres

Mineral fibres are unique category of naturally occurring or slightly treated fibres extracted from mineral deposits. Amongst its divisions are asbestos, the sole present in mother nature mineral fibre. Varieties of asbestos are serpentine (chrysotile) and amphiboles (amosite, crocidolite, tremolite, actinolite, and anthophyllite) preferred in earlier times, because of their excessive warm resistance and insulation, these fibres have created continuously registered threats to human health [14]. Another important group are ceramic fibres, such as silicon carbide, boron carbide, aluminium oxide, glass fibres, including glass wool and quartz. They are employed in many diverse industries including aeronautics, constructions, and the automotive field due to their resistance to high temperatures and tough nature. Metal fibres, such as aluminium fibres, are also noteworthy as an element with a feeling of lightness and conductivity, which is utilized in aeronautics and conductivity. In conclusion, mineral fibres form a vast group of materials with diverse features that are used in numerous industrial procedures and products as key constituents [3, 6].

2.2. Composition of Natural Fibre

The chemical composition of natural fibres exhibits significant variation depending on their specific type, with plant fibres presenting a complex structure. As elucidated in Table 1 and outlined in prior studies, plant fibres comprise a composite material that nature has skilfully created.

Table 1. Chemical composition of plant fibres [12]

Fibre	Hemicelluloses (wt.%)	Lignin (wt. %)	Cellulose (wt.%)	Moisture Content (wt.%)	Pectin (wt.%)	Waxes (wt.%)
Cereal Straw	15-31	12-20	38-45	-	8	-
Coir	0.15-0.25	40-45	32-43	8	3-4	-
Cotton	5.7	-	85-90	7.85-8.5	0-1	0.6
Abaca	-	12-13	56-63	5-10	1	-
Nettle	-	-	86	-	-	11-17
Henequen	4-8	13.1	77.6	-	-	-
Palm	-	5-12.7	70-82	11.8	-	-
Sisal	10-14	10-14	66-78	10-22	10	2
Banana	10	5	63.6	10-12	-	-
Hemp	17.9-22.4	3.7-5.7	70-74	6.2-12	0.9	0.8
Flax	18.6-20.6	2.2	71	8-12	2.3	1.7
Jute	13.6-20.4	12-13	61-71.5	12.5-13.7	0.2	0.5
Kenaf	21.5	8-13	45-57	-	3-5	-
Ramie	13.1-16.7	0.6-0.7	68.6-76.2	7.5-17	1.9	0.3

Natural fibres are referred to as lignocellulosic fibres as the main constituents are lignin and cellulose. Every component's characteristic adds to the fibre's overall characteristics. The fibres are made up of cellulose microfibrils bonded together by lignin. The fibre structure is schematically represented in Figure 3. The majority of the final cell walls are composed of α-cellulose, whose molecular chains run almost parallel to the fibre axis.

However, the majority of the hemicellulose and lignin are found in the space between neighbouring boring cells, where they serve as the middle lamella's cementing agent and promote robust lateral adhesion [9].

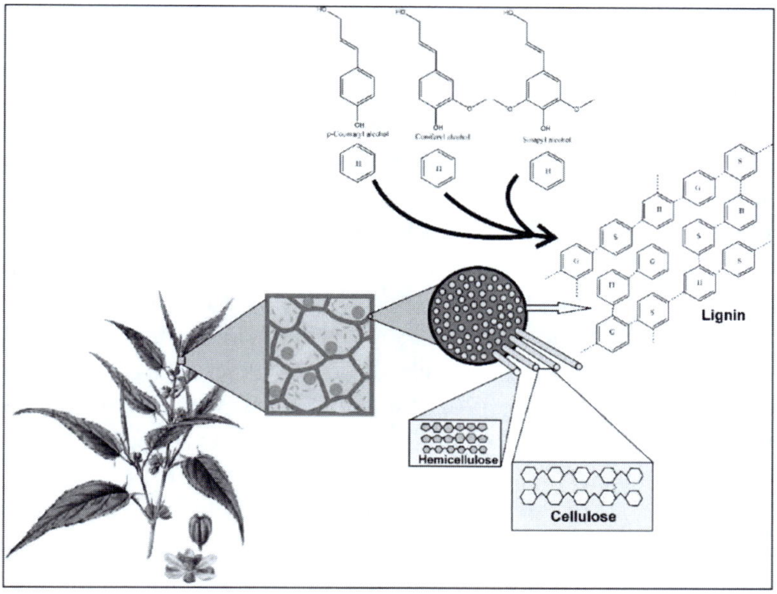

Figure 3. Structure of lignocellulosic fibre showing its hierarchical composite-like structure from stem to fibre with cellulose, hemicellulose, and lignin units.

2.3. Chemical Modification of Natural Fibre Surface

The interfacial adhesion between natural fibres and polymeric matrices is crucial for enhancing the mechanical properties of composite materials. Natural fibres possess inherent strength but are characterized by brittleness and limited extension at break due to the arrangement of cellulose molecules and the formation of hydrogen bonds. The hydrophilic nature of these fibres further complicates their integration with hydrophobic polymer matrices during composite fabrication processes [3]. In view of the above challenge, the surface modification of natural fibres is essential to achieve a better interface for bonding. Depending on the nature of the polymeric matrix, certain chemical methods with varying merits and limitations have been tried with alkali treatment, cyanoethylation, silane treatment, esterification, acetylation, and grafting approaches. The success of surface-treated composites in achieving superior composite performance also depends on the method of chemical treatment employed, concentration of treatment, and length of treatment. Therefore, the surface treatment method must be carefully selected and optimized to ensure optimal interfacial bonding performance prior to use in natural fibre reinforced polymer composite applications [15].

2.3.1. Alkali Treatment

The alkali treatment is a prevalent technique for altering the natural fibre's surface to enhance absorbency of adhesive moisture [16]. Polymorphous forms of native cellulose could be altered by both thermal and chemical process. When cellulose fibre is treated with alkali, dissolution is done by the cellulose as well as the crystalline form in Figure 6. The cellulose fibre alkali-treated swelling reaction causes the cellulose's molecular crystal to be dimmed. Some researchers have investigated natural fibre alkali treatments for potential improvement of mechanical properties and high temperatures. According to Lassoued et al., (2021) studied the date palm fibres had a tensile strength of 242.2 MPa after being treated with NaOH, as opposed to 175 MPa for untreated fibres [17]. The NaOH treatment influenced the degree of swelling to form alkali cellulose. Sodium ion (Na^+) is known to have the right diameter, which enables it to expand the tiniest pores among the lattice planes. of cellulose and penetrate them resulting in higher amount of swelling [18]. After being treated with an alkali, the hydroxyl groups (-OH) in cellulose are changed to ONa- groups, and the associated Na-ions are then eliminated by rinsing with water. The schematic interaction of NaOH with cellulose of natural fibre is shown in Figure 4.

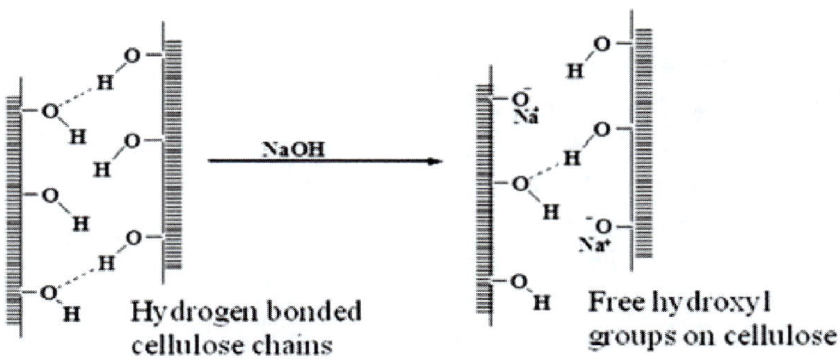

Figure 4. Schematic illustration of sodium hydroxide treatment on cellulose.

2.3.2. Benzoylation Treatment

Chemical treatments are essential in the quest to improve mechanical characteristics of composites enhanced with natural fibres. Benzoylation treatment, a post-alkali pre-treatment process, has been gaining widespread attention owing to its ability to decrease natural fibre wettability and increase thermal stability of the fibre. In a study carried out by Safri et al., (2018) they explored how treating sugar palm fibre, with benzoylation affects its properties [19]. Furthermore, the composites treated with benzoyl showed an improvement, in storage modulus pointing to a connection between the fibre and matrix. The findings indicated that benzoylation led to changes in the loss modulus of the composites indicating a reduction in energy dissipation, during deformation. According to a proposal applying benzoylation treatment may improve the characteristics of composites containing fibres. In research conducted by Izwan, (2021) it was discovered that benzoylation treatment resulted in a reduction, in fibre diameter suggesting the removal of a layer, from the surface of the fibre. As a consequence of the removal of the amorphous layers, i.e., the fibre and polymer matrix interfacial adhesion increase the composite showed the highest tensile strength and modulus, which were 173.99 MPa and 6.64 GPa, respectively [20]. Furthermore, Madhu et al., (2019) studied the mechanical properties of benzoyl-treated agave americana fibre-reinforced composites. The tensile strength enhancement of composites is found to be 70% in comparison with raw fibre reinforced composite [21]. The schematic interaction between benzoyl and natural fibre surface is shown in Figure 5.

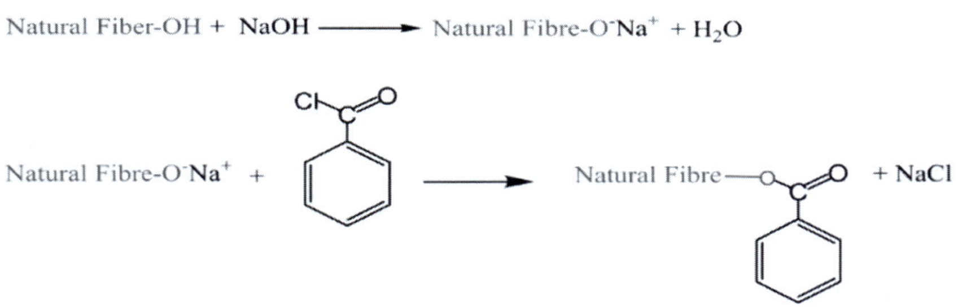

Figure 5. Schematic illustration of benzoylation of cellulose.

2.3.3. Bleaching Treatment

In this context, alkali treatment and bleaching processes appear a viable method of enhancing the characteristics of natural fibres, since these can get rid of residual lignin and hemicellulose. Kumneadklang et al., (2019), exemplify the approach as sodium hydroxide pellets are used to alkalinize, or treat the cellulose, of the palm fibres and hydrogen peroxide is later applied for bleaching. As a result, the quality of the fibre improves, as well as surface morphology and crystallinity. Thus, the above method can be considered to improve natural fibres and enhance their performance properties, while identifying an opportunity for the creation of more sustainable and ecological materials [22]. Johar et al. conducted their research on the impact of alkali treatment and bleaching on rice husk fibres outcomes. The crystallinity index and thermal stability were more influenced when a mild treatment of NaOH was used, but upon bleaching, the cellulose content enhanced, accompanied by the crystallinity index. Thus, pulping was perceived with the enhanced thermal stability that was greater for pulps derived from alkaline treatment. The process of bleaching resulted in the consequent colour changes of the fibre, which can also suggest the centre of non-cellulosic components and surface layer [23].

2.3.4. Acetylation Treatment

The use of the chemical process "acetylation treatment," a process that involves adding acetyl groups to natural fibres, is a common procedure to improve their mechanical properties and durability in composite materials. Esterification, a process that replaces the hydroxyl group (-OH) in natural fibres with an acetyl group (CH_3CO), is utilized on fibres with acetic anhydride and acetic acid. The interaction between benzoyl and the natural fibre surface is illustrated in a schematic Figure 6. Senthilraja et al., (2020) investigated the durability and mechanical properties concerning the impact of acetylation on kenaf fibre-reinforced vinyl-ester composites. Palm fibres were immersed in vinyl ester dissolved in varied concentrations and sun-cured in a given specification. This study indicated that the behaviour of Acetylation treated palm fibres has highly developed the composite mechanical mannerism and durability by reducing the moisture absorption and improving the fibre-matrix and fibre-fibres interfacial bond strength. It is evident that the process has drastically improved the properties of the composites; thus, Acetylation treatment is good to do enhancing performance of NFRC (natural fibre reinforced composite) [24]. Fitch-Vargas et al., (2019) researched the properties of a starch-based bio composite reinforced with acetylated sugarcane fibres. Sugarcane fibres were treated with NaOH and subsequent bleaching prior to being acetylated in acetic acid. Treated fibres actually presented with superior mechanical properties, lower water affinity, and higher thermal stability in comparison to the untreated fibres. Fourier

Transform Infrared analysis confirmed that the presence of hydrophobic acetyl groups in the sugarcane fibres was responsible for this improvement [25]. In conclusion, based on the studies by Senthilraja et al., (2011) acetylation treatment was seen to have multiple advantageous effects on the natural fibres and their composites. Hence, the studies all emphasize the use of the acetylation treatment as a useful and beneficial technique for enhancing natural fibre-reinforced composites performance and long-term viability [26].

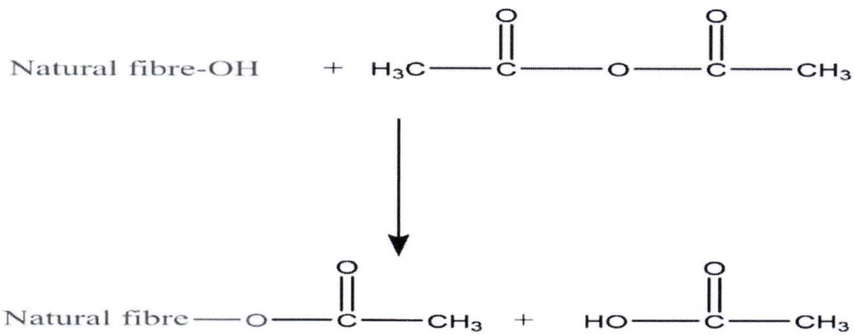

Figure 6. Schematic illustration of acylation treatment on cellulose.

2.3.5. Silane Treatment

Silane treatment is an essential coupling agent in composite material to enhance the interfacial adhesion between natural fibres and polymer matrices. This treatment involves filling of micropores located on the natural fibre's surface which increase bonding strength at the fibre-matrix interface. Figure 7 below shows the chemical reaction for silane treatment. Silane molecule covalently connects to the hydroxyl groups at the fibre surface and causes a change in the surface chemistry of the fibre. Another study conducted by Asim et al., (2017) investigated the impact of silane on the fibres from alkali-treated kenaf and pineapple leaves. The SEM images showed that the treated fibres had a smooth surface while the untreated ones were rough, which was due to the removal of lignin and hemicellulose by the alkali, and the subsequent silane completion coating the free spaces, thereby smoothing the fibres. From the study, it was concluded that the silane treatment was effective in improving the hydrophilicity of pineapple leaf and kenaf fibres and that the 2% silane and 6% NaOH treatment was better [27]. Orue studied by the same principle of sequential covalent attachment of silanes to the surface, explored a way of how a minute sisal lactic fibre composite may yield a perfect smooth and well-bristled surface contact. Microscopic inspection of the fibre structure after treatment showed less fibre density (which was assigned to contaminants) due to the decline of lignin and hemicellulose which was in contact with fibre surface. Also, the results showing adhesions between elements prove that the treatment with silane may have changed surface properties [28]. Liu et al., (2019) studied in this case, corn stalk was immersed in different concentrations of the silane coupling agent, cleaned of it using a solvent, and dried to prepare for further composite preparation. Since the fibre lost lignin and hemicellulose on the surface due to the treatment, the amount of absorbed water by it significantly decreased, which influenced the hydrophobicity and bonding properties with the polymer matrix [29].

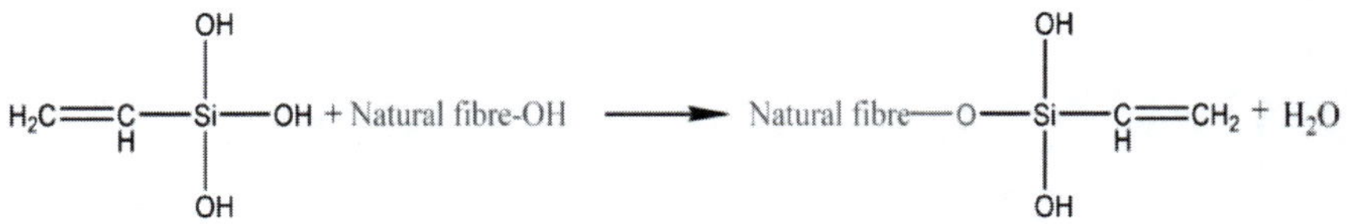

Figure 7. Schematic illustration of silane treatment on cellulose.

2.3.6. Permanganate Treatment

The potassium permanganate (KMnO4) solution in acetone used in permanganate treatment is a commonly used technique to improve the interfacial properties of composites bound with natural fibres. The production of MnO^{3-} ions during this treatment results in the synthesis of cellulose radicals, which are then used to initiate graft copolymerization with the help of extremely reactive Mn^{3+} ions. research on the effects of permanganate treatment on various fibres have been conducted; this research usually involves an alkaline pre-treatment phase [30-31]. Paul et al., (2011) submerged sisal fibres that had been subjected to an alkaline treatment in KMnO4 solutions containing acetone at concentrations ranging from 0.033% to 0.125% for a duration of one minute. The results of permanganate treatment showed that fibres' hydrophilic tendency was reduced, which in turn reduced the amount of water that fibre-reinforced composites absorbed. It's interesting to note that when KMnO4 concentrations increased, the hydrophilic propensity showed a decreasing trend. Nevertheless, at greater concentrations—1%, in particular—cellulosic fibres degraded and polar groups formed between the matrix and fibre [31]. A schematic exchange between benzoyl and natural fibre surface is shown in Figure 8.

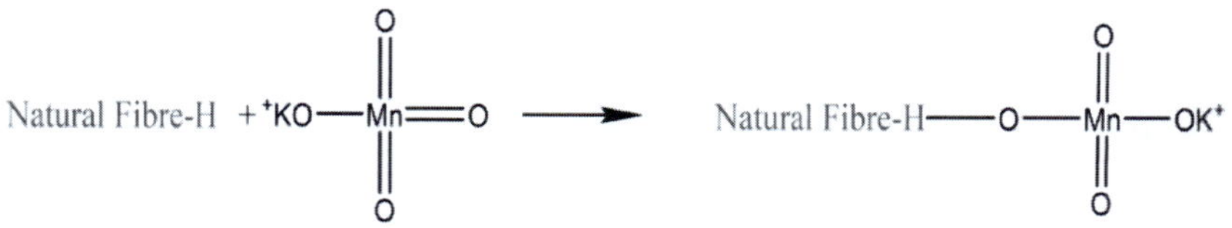

Figure 8. Schematic illustration of permanganate treatment on cellulose.

2.3.7. Peroxide Treatment

Natural fibre surface modifications require organic peroxides to enhance the performance characteristics of composite materials. Benzoyl peroxide and dicumyl peroxide are the most well-known organic peroxide initiators that initiate the above free radical reactions. These processes are catalysed by peroxides, which enhance the above mechanical characteristics by assisting the polymer matrix in bonding to cellulose fibres. Fibres usually go through an alkali pre-treatment before being coated with BP or DCP that has been dissolved in acetone solution in peroxide treatment procedures. Research has shown that higher temperatures accelerate the breakdown of peroxide by improving the oil palm fibre-reinforced phenolic composites treated with peroxide's tensile stress capability [32]. Peroxide concentration has been shown by Joseph and Thomas (1996) to be significant in short sisal fibre-modified polyethylene composites. Tensile strength values increased as peroxide concentration increased up to a certain point (4% for DCP and 6% for BP), after which a plateau effect was noted [33]. Moreover, it was discovered that after being treated with peroxide, the hydrophilicity of the fibres decreased, but the tensile qualities of the resulting composites improved at the same time. A schematic interaction between benzoyl and natural fibre surface is given in Figure 9.

$$\text{RO-OR} \longrightarrow 2\text{RO}^{\cdot}$$

$$\text{Natural Fibre-H} + \text{RO}^{\cdot} \longrightarrow \text{Natural Fibre} + \text{ROH}$$

Figure 9. Schematic illustration of peroxide treatment on cellulose.

2.3.8. Physical Surface Treatment Methods

Alternatively, the formation of fibrous-based polymer nanocomposites can be achieved using surface modification techniques mentioned above as cold plasma treatment and corona discharge. Through them, it is possible to enhance the energy and surface structure of the fibres, which positively affects their compatibility with polymer matrices. The studies have shown that in the process of oxygen plasma treatment of rayon fibres, the oxygen/carbon ratio increases within the time and consequently the polar as well as the total components of free surface energy [34]. However, the same results with cellulose fibres and the gradual increase in corona current intensity have been documented by Belgacem et al., (2007) who suggest similar uniformity for the other fibre types [35]. Yamamoto et al., (2017) showed more moistened the nature of wood surfaces in corona treatment that caused high amounts of corona. These results allow confirming the effectiveness of activation methods in promoting interfacial bonding [36]. Other findings indicate that the stimulation of the wood surface dramatically increased the amount of aldehyde groups, giving more insight into the mechanism responsible for the modification of the surface as a result of the treatment [37]. Apart from surface treatments, other agents have been evaluated for their performance as coupling agents or surface modifier to improve fibre-matrix adhesion. Anhydrides, maleate polymers, isocyanates, and alkoxysilanes are some of the common agents. The adhesion strengthening effect was observed when incorporated with cellulose fibres before matrix integration, including malleated polypropylene (MaPP) and malleated polyethylene (MaPE). As an outcome of the above, studies show improvements of between 40% and 80%, which demonstrates the efficacy of the addition of coupling agents that enhance the mechanical properties of likes of composites [38-39].

3. Polyamide (Nylon)

Poly aliphatic amides are a type of amide that comprises nylon 6, nylon 6, nylon 6, 10, and so on, whereas Kevlar is an aromatic polyamide. Aromatic polyamides have great strength and impact capabilities; hence they are utilised to make bulletproof jackets. Nylon compounds are also used to make rope, nets, shoes, and fabrics. Birch et al. (2015) depicted Nylon also has superior mechanical properties, greater resistance to corrosive substances, and visually pleasing surfaces. Due to the expensive cost of nylon and the challenges of creating composites without significant thermal degradation of lignocellulosic fillers, there is a dearth of knowledge on lignocellulosic filler reinforced nylon composites [40]. According to Graff et al. (2005), nylon composites are replacing more traditional materials such as metals and thermoset plastics due to their high tensile strength and modulus, resistance to high temperatures, oils, and corrosive chemicals, low weight, appealing surface appearance, and potential for part consolidation [41]. Tajvidi et al. (2009) demonstrated the feasibility of producing nylon/natural fibre composites [42].

3.1. Mechanical and Thermal Properties of Natural Fibre-PA6 Composites

Natural fibres are an enticing alternative to the more expensive glass fibres that are commonly used to reinforce plastics because of their low cost, low density, low abrasion, and superior mechanical properties. Tormos et al. (2024) investigated polyamides (nylon 6 and nylon 6/6) and reported that these thermoplastics are used in numerous critical automobile parts, including the parts of the fuel system, cooling systems, gearbox, and engine [43]. Lee et al. (2023) investigated the thermal deterioration of natural fibre during nylon-natural fibre based composite fabrication due to high melting temperatures (above 220 °C for nylon 6 and 6/6), which reduced the mechanical properties of composites [44]. Madueke et al., (2023) examined polyamide-based natural fibre composites and discovered that the use of fibres instead of fibre mat improved the materials' properties [45].

Ozen et al. (2013) undertook a comprehensive investigation into the thermal characteristics of a novel natural fibre blend consisting of hemp, kenaf, and flax incorporated with PA6. A detailed characterization of the blends' thermal profiles was undertaken through the employment of different marine approaches namely differential scanning calorimetry, dynamic mechanical thermal analysis, and thermogravimetric analysis. The rise in tensile and flexural characteristics indicated that the natural fibres and nylon 6 formed an efficient connection. The scanning electron microscopic examination of the fracture surfaces revealed no fibre pullout [46]. The viscoelastic and thermal properties of composites consisting of polyamide 6 and a natural fibre blend with varying percentage levels (in the range of up to 20 wt.%; flax, kenaf, and hemp fibres) have been studied by Kiziltas et al., (2016) [47].

One of the outstanding contributions to the field of composite materials was made by Kiziltas (2015). The article presents a novel approach to fabrication composites using bio-based PA610/RPA 6 blends as a matrix and cellulose-based fibres as reinforcement. Moreover, the author thoroughly conducted a series of experiments, during which Kiziltas fully characterized the mechanical and thermal properties, crystallization behaviour, and thermal degradation kinetics of these polymer composites. Furthermore, the study attempted to establish connections between these properties and phase morphology, explaining the bio-based composite's performance. Below are the mechanical properties of other natural fibre reinforced polyamide composites made by several other authors in Table 2.

Table 2. Mechanical properties of natural fibre-polyamide composites

Natural fibre	Matrix	Tensile strength (MPa)	Flexural strength (MPa)	Young's modulus (MPa)	References
Banana	PA6	8.685	-------	126.98	[48]
Jute	Epoxy and PA6	31-33	83-85	------	[49]
Curaua	PA6	104 ± 1	158 ± 4	------	[50]
Pineapple leaf	PA6	1,820 ± 49	78.2 ± 2.1	----	[51]
Flax	Bio polyamide	33.4	55.7	----	[52]

Conclusion and Future Aspects

Natural fibres in composite development have opened opportunities for integrating sustainability with performance improvement. Performance and lightness have made synthetic fibres the incumbent; natural fibre composites offer another contender due to high-performance potential and the desire to be cost-effective, lighter, and more admirable regarding mechanical properties. The adoption of natural fibres with polyamide matrix has some drawbacks ranging from compatibility, thermal stability. These challenges have slowed the incorporation of natural fibre-polyamide composites in a broad share of industries. Surface treatment techniques that include alkaline treatments, chemical functionally methods have proved effective in case of natural fibres and polyamide matrix since it improves the fibre-matrix bonding and the interface combine. Moreover, novelty in optimized fibre extraction and pre-treatment techniques and continued improvement in polymer matrix formulation are determinants of the ultimate success and viability of natural fibre composites based on their performance and durability.

Moreover, the hybridization of natural fibres and synthetic fibres creates an opportunity to minimize the composite's absorption of water while enhancing their overall properties. The possibilities for addressing each of these challenges create pathways for the realization of various applications within diverse industries, including the automotive industry. The combination of development and options offers an exceptional option for natural fibres to contribute to future solutions establishing a balance between high performance and sustainability in modern engineering and manufacturing.

References

[1] Lu, N., Swan Jr, R. H., & Ferguson, I. (2012). Composition, structure, and mechanical properties of hemp fiber reinforced composite with recycled high-density polyethylene matrix. *Journal of Composite Materials,* 46(16), 1915-1924. https://doi.org/10.1177/0021998311427778.

[2] Behera, A. K., Mohanty, C., & Mohanty, A. (2022). Development and characterization of biodegradable jute reinforced sapling pot. *Journal of Natural Fibers*, 19 (14), 7738-7749. https://doi.org/10.1177/0021998311427778.

[3] Avancha, S., Behera, A. K., Sen, R., & Adhikari, B. (2020) Accelerated weathering analysis of jute reinforced cashewnut shell liquid modified soy based green composite, *SPE Polymers* 1 (2), 81-89. https://doi.org/10.1002/pls2.10020.

[4] Ho, M. P., Wang, H., Lee, J. H., Ho, C. K., Lau, K. T., Leng, J., & Hui, D. (2012). Critical factors on manufacturing processes of natural fibre composites. *Composites Part B: Engineering,* 43(8), 3549-3562. https://doi.org/10.1016/j.compositesb.2011.10.001.

[5] Liang, X., Huang, Z., Zhang, Y., Hu, H., & Liu, Z. (2013). Synthesis and properties of novel superabsorbent hydrogels with mechanically activated sugarcane bagasse and acrylic acid. *Polymer bulletin,* 70, 1781-1794. https://doi.org/10.1007/s00289-013-0921-4.

[6] Pattnaik, S. S., Mohapatra, S. K., Mohanty, C., Behera, A. K., & Tripathy, B. C. (2023). Influence of Waste Vetiver Root Fiber on Mechanical, Hydrophobicity, and Biodegradation of Soy-Based Biocomposites as Plastic Substitute. *Fibers and Polymers,* 24(1), 265-274. https://doi.org/10.1007/s12221-023-00085-z.

[7] Feng, H., Li, J., & Wang, L. (2010). Preparation of biodegradable flax shive cellulose-based superabsorbent polymer under microwave irradiation. *BioResources,* 5(3), 1484-1495. http://dx.doi.org/10.15376/biores.5.3.1484-1495.

[8] Behera, A. K., Mohanty, C. (2021). Fabrication and characterization of nano-biocomposites using natural fiber and soy resin. *Materials Today: Proceedings*, 47(5), 1183-1188. doi: 10.1016/j.matpr.2021.03.408.

[9] Behera, A. K., Sen, R., & Adhikari, B. (2022). Environment-friendly fully biodegradable jute-poly (vinyl alcohol) modified soy composite development as plastic substitute. *Journal of Natural Fibers*, 19 (3), 905-914. doi: 10.1080/15440478.2020.1764440.

[10] May-Pat, A., Valadez-González, A., & Herrera-Franco, P. J. (2013). Effect of fiber surface treatments on the essential work of fracture of HDPE-continuous henequen fiber-reinforced composites. *Polymer Testing,* 32(6), 1114-1122.

[11] Chen, W., Yu, H., Liu, Y., Chen, P., Zhang, M., & Hai, Y. (2011). Individualization of cellulose nanofibers from wood using high-intensity ultrasonication combined

with chemical pretreatments. *Carbohydratepolymers*, 83(4),1804-1811. https://doi.org/10.1016/j.carbpol.2010.10.040.

[12] Hargitai, H., & Rácz, I. (2000). Influence of water on properties of cellulosic fibre reinforced polypropylene composites. *International Journal of Polymeric Materials and Polymeric Biomaterials*, *47*(4), 667-674. https://doi.org/10.1080/00914030008031321.

[13] Ellis, W. D., & O'dell, J. L. (1999). Wood–polymer composites made with acrylic monomers, isocyanate, and maleic anhydride. *Journal of Applied Polymer Science*, 73(12), 2493-2505. https://doi.org/10.1002/(SICI)1097-4628(19990919)73:12%3C2493.

[14] Behera, A. K., Mohanty, C., & Das, N. (2021). Synthetic glass and jute fabric reinforced soy-based biocomposites: Development and characterization. *Polymers and Polymer Composites*, 29 (9), S678-S687. doi:10.1177/09673911211020609.

[15] Mitra, B. C., Basak, R. K., & Sarkar, M. (1998). Studies on jute-reinforced composites, its limitations, and some solutions through chemical modifications of fibers. *Journal of Applied Polymer Science*, 67(6), 1093-1100. https://doi.org/10.1002/(SICI)1097-4628(19980207)67:6%3C1093.

[16] Patil, Y. P., Gajre, B., Dusane, D., Chavan, S., & Mishra, S. (2000). Effect of maleic anhydride treatment on steam and water absorption of wood polymer composites prepared from wheat straw, cane bagasse, and teak wood sawdust using novolac as matrix. *Journal of Applied PolymerScience*, 77(13), 2963-2967. https://doi.org/10.1002/10974628(20000923)77.

[17] Lassoued, M., Crispino, F., & Loranger, E. (2021). Design and synthesis of transparent and flexible nanofibrillated cellulose films to replace petroleum-based polymers. *Carbohydrate Polymers*, 254, 117411 https://doi.org/10.1016/j.carbpol.2020.117411.

[18] Joseph, K., Thomas, S., & Pavithran, C. (1996). Effect of chemical treatment on the tensile properties of short sisal fibre-reinforced polyethylene composites. *Polymer*, 37(23), 5139-5149. https://doi.org/10.1016/0032-3861(96)00144-9.

[19] Safri, S. N. A., Sultan, M. T. H., Jawaid, M., & Jayakrishna, K. (2018). Impact behaviour of hybrid composites for structural applications: A review. *Composites Part B: Engineering*, 133, 112-121. https://doi.org/10.1016/j.compositesb.2017.09.008.

[20] Izwan, S. M., Sapuan, S. M., Zuhri, M. Y. M., Mohamed, A. R., & Ilyas, R. A. (2021). A comprehensive review of natural fiber reinforced polymer biocomposites and their applications. *Design for Sustainability*, 287-305. https://doi.org/10.1016/B978-0-12-819482-9.00019-8.

[21] Madhu, P., Sanjay, M. R., Senthamaraikannan, P., Pradeep, S., Saravanakumar, S. S., & Yogesha, B. (2019). A review on synthesis and characterization of commercially available natural fibers: Part II. *Journal of Natural Fibers*, 16(1), 25-36. https://doi.org/10.1080/15440478.2017.1379045.

[22] Kumneadklang, S., Sompong, O., & Larpkiattaworn, S. (2019). Characterization of cellulose fiber isolated from oil palm frond biomass. *Materials Today: Proceedings*, 17, 1995-2001. https://doi.org/10.1016/j.matpr.2019.06.247.

[23] Johar, M., Chong, W. W. F., & Wong, K. J. (2023). Moisture Absorption and Tensile Behaviour of Hybrid Carbon/Flax Composites. *Fibers and Polymers*, 24(5), 1799-1810. https://doi.org/10.1007/s12221-023-00157-0.

[24] Jiang, W., Li, L., Pan, J., Senthil, R. A., Jin, X., Cai, J., Wang, J., & Liu, X. (2019). Hollow-tubular porous carbon derived from cotton with high productivity for enhanced performance supercapacitor. *Journal of Power Sources*, 438, 226936. https://doi.org/10.1016/j.jpowsour.2019.226936.

[25] Fitch-Vargas, P. R., Camacho-Hernández, I. L., Martínez-Bustos, F., Islas-Rubio, A. R., Carrillo-Cañedo, K. I., Calderón-Castro, A., ... & Aguilar-Palazuelos, E. (2019). Mechanical, physical and microstructural properties of acetylated starch-based biocomposites reinforced with acetylated sugarcane fiber. *Carbohydrate Polymers*, 219, 378-386. https://doi.org/10.1016/j.carbpol.2019.05.043.

[26] Senthilraja, P., Kathiresan, K., & Saravanakumar, K. (2011). Comparative analysis of bioethanol production by different strains of immobilized marine yeast. *Journal of Yeast and Fungal Research*, 2(8), 113-116.

[27] Asim, M., Jawaid, M., Abdan, K. et al. Effect of pineapple leaf fibre and kenaf fibre treatment on mechanical performance of phenolic hybrid composites. *Fibers and Polymers*, 18, 940–947 (2017). https://doi.org/10.1007/s12221-017-1236-0.

[28] Orue, A., Jauregi, A., Unsuain, U., Labidi, Eceiza, J. A., Arbelaiz, A. (2016). The effect of alkaline and silane treatments on mechanical properties and breakage of sisal fibers and poly(lactic acid)/sisal fiber composites. *Composites Part A: Applied Science and Manufacturing*, 84, 186-195. https://doi.org/10.1016/j.compositesa.2016.01.021.

[29] Liu, Y., Xie, J., Wu, N., Wang, L., Ma, Y., & Tong, J. (2019). Influence of silane treatment on the mechanical, tribological and morphological properties of corn stalk fiber reinforced polymer composites. *Tribology International*, 131, 398-405. https://doi.org/10.1016/j.triboint.2018.11.004.

[30] Abdullah, N. A., Sainorudin, M. H., Rani, M. S. A., Mohammad, M., Abd Kadir, N. H., & Asim, N. (2021). Structure and thermal properties of microcrystalline cellulose extracted from coconut husk fiber. *Polimery*, 66(3), 187-192. https://doi.org/10.14314/polimery.2021.3.4.

[31] Mohammed, A. A., Bachtiar, D., Rejab, M. R. M., Jiang, X. X., Abas, F. O., Abass, R. U., Hasany, S. F., & Siregar, J. P. (2018). Effects of $KMnO_4$ treatment on the flexural, impact, and thermal properties of sugar palm fiber-reinforced thermoplastic polyurethane composites. *Journal of the Minerals, Metals & Materials Society*, 70, 1326-1330. doi: 10.1007/s11837-018-2869-1.

[32] Benkaddour, A., Demir, E. C., Jankovic, N. C., Kim, C. I., McDermott, M. T., & Ayranci, C. (2022). A hydrophobic coating on cellulose nanocrystals improves the mechanical properties of polyamide-6 nanocomposites. *Journal of Composite Materials*, *56*(11), 1775-1788. https://doi.org/10.1177/00219983221075418.

[33] Azwa, Z. N., Yousif, B. F., Manalo, A. C., & Karunasena, W. (2013). A review on the degradability of polymeric composites based on natural fibres. *Materials & Design*, 47, 424-442. https://doi.org/10.1016/j.matdes.2012.11.025.

[34] Souza, S. F., Ferreira, M., Sain, M., Ferreira, M. Z., Pupo, H. F., Cherian, B. M., & Leão, A. L. (2015). The use of curaua fibers as reinforcements in composites. In

Biofiber reinforcements in composite materials (pp. 700-720). Woodhead Publishing. https://doi.org/10.1533/9781782421276.5.700.

[35] Belgacem, M. N., & Gandini, A. (2005). Surface modification of cellulose fibres. *Polímeros*, 15 (2). https://doi.org/10.1590/S0104-14282005000200010.

[36] Yamamoto, A., Kymäläinen, M., Lindroos, T., Rohuma, A., Sokka, K., & Rautkari, L. (2017). Surface activation of wood by corona treatment and NaOH soaking for improved bond performance in plywood. *Bio Resources*,12(4), 9198-9211. doi: 10.15376/biores.12.4.9198-9211.

[37] Ferra, J. M. M., Ohlmeyer, M., Mendes, A. M., Costa, M. R. N., Carvalho, L. H., & Magalhães, F. D. (2011). Evaluation of urea-formaldehyde adhesives performanceby recently developed mechanical tests. *International Journal of Adhesion and Adhesives*, 31(3), 127-134. doi:10.1016/j.ijadhadh.2010.11.013.

[38] Costa, N., Pereira, J., Martins, J., Ferra, J., Cruz, P., Magalhães, F., Mendes, A., & Carvalho, L. (2012). Alternative to latent catalysts for curing UF resins used in the production of low formaldehyde emission wood-based panels. *International Journal of Adhesion and Adhesives*, 33, 56-60. doi: 10.1016/j.ijadhadh. 2011.11.003.

[39] Abdelmouleh, M., Boufi, S., Belgacem, M. N., & Dufresne, A. (2007). Short natural-fibre reinforced polyethylene and natural rubber composites: Effect of silane coupling agents and fibres loading. *Composites science and technology*, 67(7-8), 1627-1639. https://doi.org/10.1016/j.compscitech.2006.07.003.

[40] Birch, A., Dal Castel, C., Kiziltas, A., Mielewski, D., & Simon, L. (2015). Development of cost effective and sustainable polyamide blends for automotive applications. In *SPE Automotive Composites Conference & Exhibition* (pp. 1-10).

[41] Arsad, A., Suradi, N. L., Rahmat, A. R., & Danlami, J. M. (2013). The influence of kenaf fiber as reinforcement on recycled polypropylene/recycled polyamide-6 composites. *International Journal of Plastics Technology*, 17, 149-162. https://doi.org/10.1007/s12588-013-9055-7.

[42] Tajvidi, M., Feizmand, M. (2009). Effect of cellulose fiber reinforcement on the temperature dependent mechanical performance of nylon 6. *Journal of Reinforced Plastic and Composites*, 28(22), 2781–90. https://doi.org/10.1177/0731684408093875.

[43] Tormos, B., Bermudez, V., Balaguer, A., & Gimenez, E. (2024). Compatibility study of polyamide (PA6) with lubricant bases for electric vehicle applications. *Lubricants*, 12(2), 54. https://doi.org/10.3390/lubricants12020054.

[44] Lee, J. A., Kim, J. Y., Ahn, J. H., Ahn, Y. J., Lee, S. Y. (2023). Current advancements in the bio-based production of polyamides. Trends in Chemistry, 5(12), 873-891. https://doi.org/10.1016/j.trechm.2023.10.001.

[45] Madueke, C. I., Mbah, O. M., & Umunakwe, R. (2023). A review on the limitations of natural fibres and natural fibre composites with emphasis on tensile strength using coir as a case study. *Polymer Bulletin*, 80, 3489–3506. https://doi.org/10.1007/s00289-022-04241-y.

[46] Ozen, E., Kiziltas, A., Kiziltas, E. E., & Gardner, D. J. (2013). Natural fiber blend—nylon 6 composites. *Polymer Composites*, 34(4), 544-553. https://doi.org/10.1002/pc.22463.

[47] Kiziltas, E. E., Yang, H. S., Kiziltas, A., Boran, S., Ozen, E., & Gardner, D. J. (2016). Thermal analysis of polyamide 6 composites filled by natural fiber blend. *Bio Resources*, 11(2), 4758-4769. https://dx.doi.org/0.

[48] Singh, R., Kumar, R., & Ranjan, N. (2019). Sustainability of recycled ABS and PA6 by banana fiber reinforcement: thermal, mechanical and morphological properties. *Journal of The Institution of Engineers (India): Series C*, 100, 351-360. https://doi.org/10.1007/s40032-017-0435-1.

[49] Thitithanasarn, S., Yamada, K., Ishiaku, U. S., & Hamada, H. (2012). The effect of curative concentration on thermal and mechanical properties of flexible epoxy coated jute fabric reinforced polyamide 6 composites. *Open Journal of Composite Materials*, *2*(04), 133. http://dx.doi.org/10.4236/ojcm.2012.24016.

[50] Santos, P. A., Spinacé, M. A., Fermoselli, K. K., & De Paoli, M. A. (2007). Polyamide-6/vegetal fiber composite prepared by extrusion and injection molding. *Composites Part A: Applied Science and Manufacturing*, 38(12), 2404-2411. https://doi.org/10.1016/j.compositesa.2007.08.011.

[51] Panyasart, K., Chaiyut, N., Amornsakchai, T., & Santawitee, O. (2014). Effect of surface treatment on the properties of pineapple leaf fibers reinforced polyamide 6 composites. *Energy Procedia*, 56, 406-413. https://doi.org/10.1016/j.egypro.2014.07.173.

[52] Kuciel, S., Kozniar, P., & Liber-Knec, A. (2012). Polyamides from renewable sources as matrices of short fiber reinforced biocomposites. *Polimery*, *57*(9), 627-634. doi: 10.14314/polimery.2012.627.

Chapter 4

The Mechanical Properties of Thermoplastic Polyurethane Clay Composites

Shruti Pandey*

Department of Civil Engineering, Indian Institute of Science Bangalore, Bangalore, India

Abstract

Thermoplastic polyurethane (TPU) is a block copolymer known for its distinctive structure comprising alternating hard and soft segments. Adding small amounts of inorganic fillers in the polymer matrix alters the composition and mechanical properties of the PU nanocomposites. The actual improvements in the thermal, physical, gas barrier and mechanical properties of polymer-clay nanocomposites are undeniably contingent on the extent of dispersion of the filler clay particles within the polymer matrix. The TEM images show good 30B and NK75 dispersion during moderate agglomeration in nano talc. The XRD pattern comprises distinct peaks indicating a certain level of crystallinity in the polymer matrix induced by the presence of the filler particles. A slight decrease in degradation temperature is observed, but the nanocomposites are thermally stable up to 300^0C.

The thermoplastic polyurethane (TPU) nanocomposites, fabricated through melt extrusion using diverse concentrations of talc and organically modified nanoclay as filler in the ester-type PU, outperformed the mechanical properties of pure TPU. The stiffness values exhibited a continuous increase with escalating nanoclay concentration, highlighting the superior mechanical response of the nanocomposites. The polyurethane nanocomposites prepared using

* Corresponding Author's Email: shrupandey28@gmail.com

In: Mechanical and Thermal Properties of Engineering …
Editors: Senthilkumar Krishnasamy, Thitinun Ungtrakul, Sabarish Radoor et al.
ISBN: 979-8-89530-460-0

NK75 clay showed better performance when compared to 30B and nano talc, giving further credence to improvement in properties due to better interaction between the filler particles and polymer matrix. The enhanced structural and mechanical properties of these nanocomposites position them for various applications across different industries.

Keywords: clay, composites, structure, melt blending, mechanical properties

1. Introduction

Composites have significantly influenced both research and industrial applications. In 1986, the advent of polymer clay hybrids marked a significant development at Toyota Central Research and Development Laboratories, with the creation of pioneering nylon-6-clay hybrid (NCH). Subsequent modifications involved the incorporation of diverse materials such as polystyrene, polypropylene, rubber, polyurethanes, polyesters, nitrile, polyimides, and polysiloxanes, resulting in a range of polymer-clay composites tailored for diverse uses in industry and engineering applications. Notably, monolayer clay constitutes around 1–15 vol% of NCH, exhibiting a doubled tensile elastic modulus compared to nylon 6, accompanied by an increased heat distortion temperature of 145°C from 65°C.

This exceptional performance of NCH is attributed to the ionic interaction between the layers of silicates present in the clay and nylon 6. The improvements in properties are linked to the large surface area and the high aspect ratio of the filler particles, with phyllosilicates, particularly montmorillonite (Mt), being utilized for nanocomposite preparation. The layered structure of Mt, held together by electrostatic forces and van der Waals forces, contributes to its widespread application in nanocomposites.

The synthesis of polymer–clay materials involve the combining of a polymer with natural or synthetic clay, where the clay particles are the fillers, that modify the structure enhancing the thermal and physical properties, introduces tortuosity that enhances gas barrier properties and may introduce certain amount of crystallinity enhancing the mechanical properties of the polymer. The detailed analysis of clay minerals necessitates X-ray diffraction, given their exceptionally fine nature. Clays are fine-grained natural soils that contain clay minerals, which are primarily hydrous aluminum phyllosilicates. Clay minerals have different chemical formulas, depending on the type and amount of aluminum, silicon, and other elements in their structure. These variations affect the chemical properties like the ability to exchange cations, decompose; and physical properties of clays, e.g., their ability to deform, fracture etc. and be used for various purposes.

In this chapter, we will detail the mechanical behavior of Polyurethane clay composites.

2. Polyurethane

Polyurethanes (PUs) represent a distinctive category of polymeric materials that diverge significantly from other plastic types. Their versatility allows incorporation into various applications, including liquid coatings, paints, elastomers, elastic fibers, insulators, integral skins, and foams. Originating from the pioneering work of German Professor Dr. Otto Bayer and colleagues, PUs evolved through the di-isocyanate poly-addition technique, leading to the establishment of the PU industry in 1937.

Initially developed as a substitute for rubber during World War II, PUs gained prominence due to their adaptability and capacity to replace scarce materials. One of the earliest applications of polyurethane (PU) coatings was to impregnate paper with them, creating a gas and corrosion-resistant material that could be used for making protective clothing. PU coatings also proved to be effective in providing chemical resistance to various substrates, such as metal, wood, and masonry, and were used as protective and decorative coatings for these surfaces. By the mid-1950s, PU coatings had expanded their scope of use to other domains, such as elastomers, coatings, rigid foams, and adhesives. Elastomers are rubber-like materials that can stretch and recover their shape, and PU coatings enhanced their mechanical and thermal properties. Coatings are thin films that are applied to a surface to modify its appearance or functionality, and PU coatings offered superior durability, flexibility, and weather resistance. Rigid foams are solid materials that have a cellular structure, and PU coatings improved their insulation, strength, and fire retardancy. Adhesives bond two or more materials together, and PU coatings increased their adhesion, cohesion, and resistance to solvents and moisture. Thus, PU coatings became popular in numerous uses due to their beneficial and versatile behavior. The latter part of the 1950s witnessed the commercial availability of comfortable cushions made from flexible foams derived from cost-effective polyether polyols.

Continuous advancements in processing techniques, additives, and formulations broadened the range of PU applications. Today, polyurethanes

(PUs) are among the most widely used, adaptable, and investigated materials in the world. They have a unique ability to incorporate a metal's strength and resilience along with rubber's flexibility and elasticity that leads to their suitability for replacing materials like plastics, metals, and rubber in many industrial products. PUs can be found in numerous applications, e.g., in elastomers, foams, coatings, fibers, and adhesives, and can be tailored to have different properties and characteristics depending on the desired performance and functionality. The superior mechanical properties e.g., the modulus of elasticity, elongation/strain at break, hardness, toughness and strength, have facilitated applications in biomedicine, construction, automotive, textiles, and other industries.

The urethane group, formed by the chemical reaction of alcohol (–OH) and isocyanate (–NCO) groups, is the main repeating unit in PUs. Polyurethanes encompass diverse groups like urea, ethers, esters, and aromatic compounds, allowing for synthesis from varied sources. Consequently, PUs are classified into thermoplastic, binders, waterborne, rigid, flexible, coatings, sealants, adhesives, and elastomers based on desired properties.

Thermoplastic polyurethane (TPU) is a block copolymer known for its distinctive structure comprising alternating hard and soft segments [1]. Hard segment has semi-crystalline characteristics that result from the reaction involving diol/diisocyanate and polyisocyanate, and the soft segment which is amorphous in nature is formed by polyols [2, 3]. Notably, the melting point temperature for the hard segment is high, while the glass transition temperature for the soft segment is low as reported by Buckley et al.[4]. The hard segment serves a dual role as a cross-linking point and filler, contributing discrete melting temperatures, with the elasticity in TPU chains attributed to the rubbery state of the soft segments.

TPU possesses exceptional chemical and physical properties, making it applicable across a spectrum of industries [5, 6]. Its attributes, including chemical resistance, shock absorption, abrasion resistance, and flexibility, have fueled increased demand [7, 8, 9]. However, to further enhance its strength and stiffness, filler particles are commonly incorporated into TPU composites [10, 11, 12]. Because of their widespread availability and cost-effectiveness, clay or the layered silicates are frequently chosen as fillers materials in the polymer matrices to form composites. The incorporation of these low cost filler particles not only improves the gas barrier properties by introducing tortuosity in the block copolymer [13, 14] but also imparts additional advantageous properties, including improved mechanical properties, flame retardancy, heat resistance, and more. The incorporation of these fillers contributes comprehensively to enhancing TPU's performance across diverse industrial applications.

3. Clay

Clay minerals are a diverse group of hydrous aluminum phyllosilicates with various classifications based on their structure and origin. This section explores the major classifications of clay minerals including their properties and industrial applications and highlighting their significance in material science.

Clay minerals are primarily categorized on their layered structure formed by sheets within their composition. The major classifications include chlorite, illite, smectite and the kaolinite group [15].

3.1. Illite

Illite ($(K, H)Al_2(Si, Al)_4O_{10}(OH)_2xH_2O$), derived mainly from sedimentary rock shales. Its unique structure features aluminum hydroxide/oxide layers sandwiched between two layers of silicates, resembling a layered structure [16]. Illite finds significant application as fillers in composites [17].

3.2. Kaolinite Group

The three members: Dickite, nacrite, and kaolinite, all sharing a common empirical formula, $Al_2Si_2O_5(OH)_4$, forms the kaolinite group [18]. Despite having the same chemical formula, these three clays exhibit distinct characteristics. Kaolinite is extensively used in plastic, paint, paper, rubber, ceramics, and as fillers in various applications.

3.3. Chlorite Group

The chlorite group, including members like daphnite, cookeite, amesite, and chamosite, is comparatively large and not always considered a true clay [19]. Due to their diverse chemical and physical structures, these groups are sometimes separated from phyllosilicates.

3.4. Smectite Group

The smectite group comprises five major minerals: montmorillonite, saponite, talc, pyrophyllite, and nontronite. Montmorillonite, which has a layer structure similar to illite, finds its application in various fields such as rubber, paint, porcelain resistant to heat and acid, electrical devices, plasticizers, and even conventional cosmetic powder [20].

3.5. Broad Classification of Clay Materials

A broader classification of clay materials includes saponite, zeolite, laponite, rectorite, montmorillonite, sepiolite, hectorite, kaolinite, chlorite, and vermiculite [21]. These hydrous aluminum phyllosilicates contain various cations such as alkali metals, magnesium, and iron, with intrinsic properties influencing their applications [16].

In this chapter, we will talk about polyurethane clay composites prepared using Montmorillonite clay and nanotalc.

Montmorillonite (MMT) stands out as the predominant natural layered silicate employed in the formulation of nanocomposites. Its chemical formula is $[(Na, Ca)_{0.33}(Al, Mg)_2(Si_4O_{10})(OH)_2.nH_2O]$, which indicates a hydrated sodium calcium aluminum magnesium silicate hydroxide. The montmorillonite structure consists of two different types of layers: tetrahedral and octahedral. The silicate layers of MMT are formed by two tetrahedral sheets bonded to an octahedral sheet, which contains either aluminum or magnesium hydroxide (referred to as a 2:1 layer). These layers are stacked with a regular van der Waals spacing between them, known as the interlayer. The cations within the layers are iso-morphically substituted and hence these layers have a net negative charge. This may be understood as Al^{3+} can be replaced by Mg^{2+} in the octahedral sheet or Si^{4+} by Al^{3+} in the tetrahedral sheet. This negative charge is compensated by alkali (Na^+) or alkaline earth (Ca^{2+}, Mg^{2+}) cations in the interlayer [22].

Cloisite 30B, a montmorillonite clay of natural origin, undergoes modification with a quaternary ammonium salt as part of its purification process. It has 2.5% of magnesium and 2% of iron. Composites reported in this chapter were prepared using Cloisite 30B which has ion-exchanged with methyl tallow bis-hydroxyethyl quaternary ammonium.

Indian clay NK75, a montmorillonite (Mt) clay from India, consists of two tetrahedral silicate layers and a planar octahedral alumina layer placed between the two layers [23]. The structure is periodic with a characteristic distance. Montmorillonite clay has high surface area, aspect ratio, adsorptive properties, surface reactivity and cation exchange capacity (CEC) [24]. As nanoclay NK75 has a high iron content, which in turn enhances its compatibility with living tissues. Hence, NK75 serves as a filler material for various biomedical applications [25]. Nanocomposites were prepared using Indian clay NK75 as a filler, having a high iron content and ion-exchanged with dimethyl dihydrogenated tallow ammonium with CEC = 70 meq/100 g.

Talc primarily comprises natural rocks and minerals, predominantly magnesium silicates. Talc, with the chemical formula $Mg_3(Si_4O_{10})(OH)_2$, is a representative form and theoretically consists of 63.5% of silicon dioxide (SiO_2), 31.7% of magnesium oxide (MgO), and 4.8% of water (H_2O), rendering it a hydrophobic filler. The chemical structure of talc particles comprises a layer of magnesium-oxygen/hydroxyl octahedral, known as the brucite sheet, positioned between the two layers of silicon- oxygen tetrahedral, referred to as the silica sheets. Weak van der Waals forces held these layers together that are stacked indefinitely maintaining electrical neutrality. The distinctive structure of talc results in its tendency to form in plate-like structures [26]. The nanotalc reported for preparation of composite in this chapter exhibits a median particle size (d50) of 70 nm and BET surface area determined through N_2 absorption is 11.8 m^2/g.

4. Preparation Techniques

The chapter describes three different techniques for preparing polymer clay nanocomposites. These techniques are:

4.1. In-Situ Polymerization

This methodology is specifically devised to augment the compatibility and interaction between the polymer and clay components. The procedure involves the uniform dispersion of nanoclay particles within a solution that contains a selective monomer.

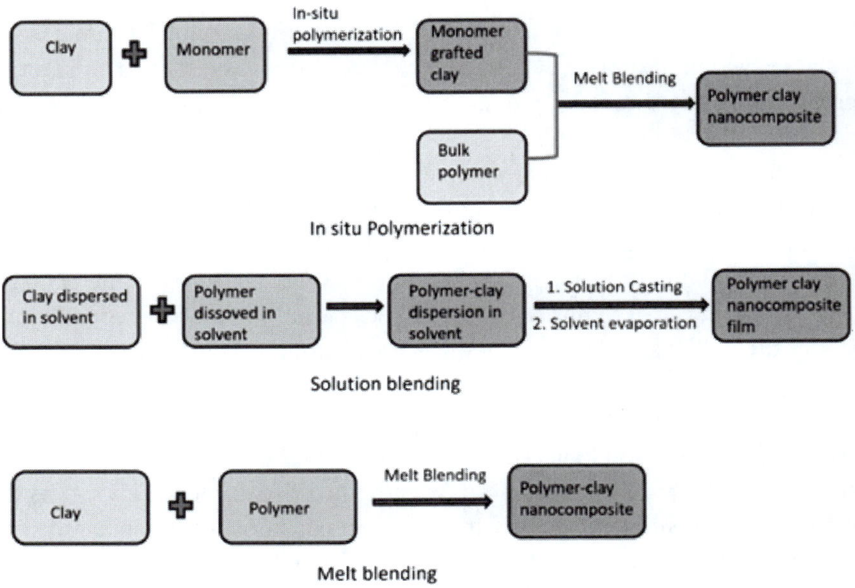

Figure 1. The three different preparation techniques of polymer clay nanocomposites.

Subsequently, this monomer undergoes polymerization through appropriate techniques. As a consequence of this, the whole clay particle becomes a grafted layer within the polymer matrix.

In the final step, this thin polymer blend, incorporating the grafted nanoclay, is introduced into the bulk polymer matrix to serve as reinforcement material. This systematic approach ensures the uniform distribution of the clay nanoparticles throughout the polymer matrix. The secondary material processing techniques like compression and extrusion, are employed as integral steps in these procedures [27].

4.2. Solution Blending

The solution blending technique represents a widely utilized and comparatively straightforward method for fabricating nanocomposites using both clay and polymer components. This process involves dispersing clay particles in a specific solvent and simultaneously dissolving the polymer in the same solvent. Then, the two solutions are mixed and the obtained mixture is homogenized for a certain period. Lastly, the homogenized mixture is poured on a flat surface or plate to enable evaporation of the solvent [28].

4.3. Melt-Blending

Given that the solution blending technique encounters challenges, particularly in achieving uniform mixing, the melt-blending technique is introduced to address these issues more effectively. This technique entails the direct incorporation of the filler particles such as clay into the polymer matrix through a melting process. Following the integration of the polymer melt, the mixture undergoes further processing through annealing. Annealing occurs at a temperature that is significantly higher than the polymer's melting point [29]. This approach enhances the overall mixing efficiency and ensures a more homogeneous dispersion of clay reinforcement within the polymer matrix. The schematic demonstrating the three methods is provided in Figure 1.

The polymer nanocomposites described in this chapter are prepared by utilizing the hot melt extrusion (HME) method of nanocomposite preparation. Holt melt extrusion is a well-established method that dates back to the 1930s. It is a continuous process in which heat and pressure are applied that results in the melting of the polymer. The polymer melt is then forced out through a small opening or orifice. Hot melt extrusion is widely used in rubber, plastic and food industries. The wide variety of plastic products such as foams, films, sheets, bags, tubes, rubbers, and pipes are produced using this technique. This technique is now a day also used in healthcare industry, particularly it finds applications in the production of medical devices and to uniformly bend the active pharmaceutical ingredients (APIs) with polymers. The proper blending of the APIs increases the bio-availability which in turn find applications in creating thermoplastic drug-eluting devices such as intraocular and subcutaneous implants, and intravaginal rings as well.

The extrusion process, known for its high production rate, involves applying heat to alter the material's viscosity, enabling it to flow through a die of the desired shape and cross-sectional area. An extruder comprises two main sections: the conveying system, responsible for proper material mixing and transportation to the die, and the die system, where the material takes on the required shape. The equipment for hot melt extrusion includes the extruder itself, assisting equipment, downstream processing apparatus, and monitoring tools. Essential components, such as feeding hoppers, barrels, the die, single or twin screws, and the screw-driving unit, form integral parts of an extruder.

Auxiliary equipment typically includes a solvent delivery pump, heating/cooling devices for the barrels, and a conveyor belt for product cooling. Monitoring devices, such as screw-speed controllers, temperature gages, extrusion torque monitors, and pressure gages, are employed for performance and product quality assessment.

The extrusion process can be theoretically understood by dividing it into four key sections: feeding through the hopper, mixing and conveying of mass, flow of the material through the die, and its exit from the die. The extruder itself consists of a stationary cylindrical barrel, often assembled in sections, with one or two rotating screws inside. The shape of the extruded product is influenced by the end-plate die connected to the barrel, while liquid or electric heaters provided on the barrels supply the necessary heat for material melting or fusion. The feed zone, featuring deeper flights or larger pitch flights, facilitates the material's easy fall into the screw so that it can be conveyed along the barrel. The transition zone, where the mixing, melting and compression of the material occur, and the material is plasticized. This zone is crucial for maintaining a consistent flow. The metering zone ensures that the material flows through the die cavity with a uniform delivery rate, reducing pulsating flow and ensuring an extrudate of uniform thickness.

In twin-screw extruders, agitator assemblies on parallel shafts can co-rotate (rotate together) or counter-rotating (rotate in opposite directions). Fully intermeshing assemblies wipe surfaces and interior chambers, ensuring a well-defined residence time distribution. Material is transferred between screws as they rotate, facilitating transportation along the extruder barrel.

To blend fillers with TPU, the authors employed a twin-screw extruder for a time duration of 10 minutes, excluding feeding time, at a temperature of 220°C and a shear rate of 100 rpm [14]. It is noteworthy that the mechanical properties of nanocomposites exhibit improvement up to approximately 5-7% wt. of filler. However, surpassing this threshold leads to agglomeration of silicate layers or the formation of aggregates, resulting in a decline in mechanical properties [30, 31, 32, 33]. Consequently, the nanocomposite formed by TPU with 4% wt. of Indian clay NK75, Cloisite 30B (C30B), and nanotalc are denoted as 'PUN', 'PUB', and 'PUT' respectively. Additionally, nanocomposites with 2%, 6%, and 10% wt. of nanoclay were reported by the authors [14, 26] to investigate the impact of nanoclay loading.

5. Structure

The actual improvements in the thermal, physical, gas barrier, and mechanical properties of polymer-clay nanocomposites are undeniably contingent on the extent of dispersion of the filler clay particles within the polymer matrix. The distribution of clay nanoparticles in the polymer matrix is influenced by two key factors: (1) the physical or chemical interaction capability between the polymer matrix and the clay material and (2) the method employed for the fabrication of polymer-clay nanocomposites.

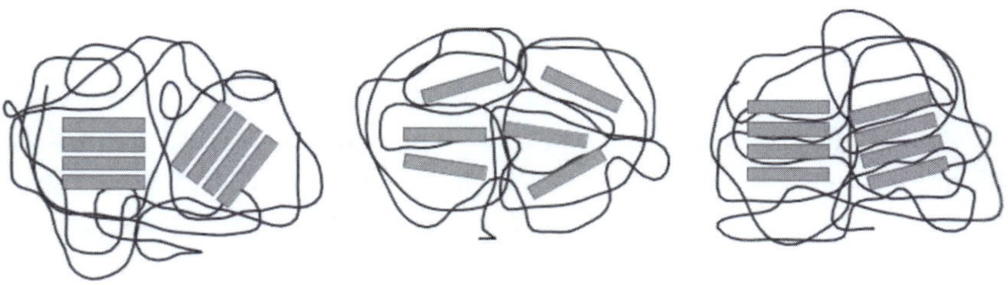

a b c

Figure 2. Schematic depicting (a) Phase separated (layered clay and polymer separate), (b) Exfoliated (delaminated layers of clay in polymer) and (c) Intercalated structure (polymer within the layered clay).

Consequently, based on the disparity in interaction affinity between the polymer matrix and the filler clay particles, three potential polymer- clay composite structures can be identified, as elucidated above and shown in Figure 2.

5.1. Phase-Separated Structure

This structure is formed when the clay material and the polymer have a very low interaction affinity, which prevents the polymer material from penetrating into the clay layers. The resultant composite structure that emerges from this process is called a microcomposite, not a nanocomposite, because it does not have nanoscale features. The microcomposite is expected to have inferior properties compared to the nanocomposite [34, 35].

5.2. Exfoliated Structure

The exfoliated/delaminated structure occurs when clay layers are completely exfoliated into individual sheets and these sheets are uniformly dispersed in the polymer phase. The fully exfoliated clay sheets when uniformly dispersed in the polymer matrix is expected to play a crucial role in enhancing the thermal, physical, gas barrier and mechanical properties of the resulting polymer-clay nanocomposite. However, complete exfoliation of the clay layers is an arduous exercise in practical situations [34, 35].

5.3. Intercalated Structure

The process of fabricating polymer-clay nanocomposites involves inserting polymer chains between the clay layers. However, if the insertion is partial, an intercalated structure is formed. In this type of structure, the polymer-clay nanocomposites have a larger interlayer spacing, which results from the partial penetration of the polymer chains into the clay layers [34, 35].

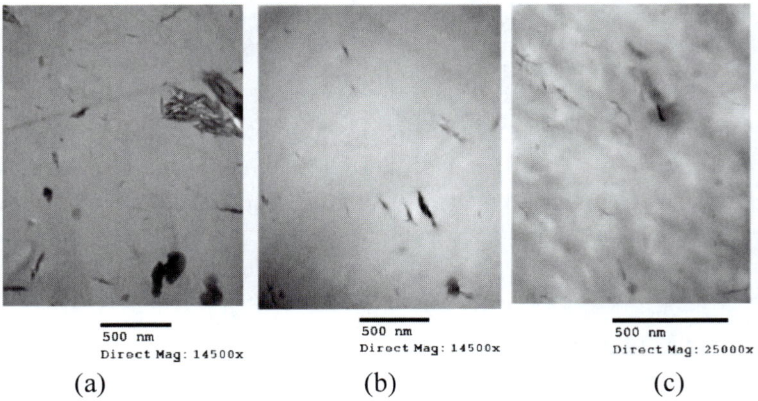

Figure 3. The bright field TEM images of (a) PU-Talc nanocomposites [26] showing less favorable dispersion, (b) PU-NK75 [14] showing uniform dispersion of Indian clay and (c) PU-C30B nanocomposites [26] depicting well dispersed layers of the C30B.

5.3.1. TEM

The bright field TEM images of the nanocomposites are depicted in Figure 3. In the PU-C30B composite, well-dispersed silicate layers of Cloisite 30B are observed within the TPU matrix. Conversely, the TEM image of PU-Talc reveals a less favorable dispersion of the filler within the PU matrix [26]. The organically modified NK75 nanoclay exhibits uniform dispersion in TPU, showcasing stacked nanoclay layers. Additionally, a high-resolution TEM image highlights the intercalated structure observed in the PU-NK75 composites [14].

5.3.2. Small Angle X-Ray Diffraction

Figure 4 depicts the XRD patterns at lower angles. The peak of Cloisite 30B is evident at $2\theta = 4.8^{o}$ (corresponding to a d-spacing of 1.85 nm). However, in the case of thermoplastic polyurethane (TPU) nanocomposites incorporating C30B, no discernible peaks are observed. This absence of peaks suggests that the silicate layers of C30B are effectively dispersed within the TPU polymer matrix. The hydroxyl groups in Closite 30B establish hydrogen bonds with the carbonyl groups in PU, fostering improved dispersion of C30B within the PU matrix [25].

For nanocomposites incorporating Talc, the intensity of the peak at $2\theta = 9.3^{o}$ (consistent with pure Talc) is diminished, while the d-spacing remains unaltered. This reduction in intensity may indicate slight intercalation and dispersion of nano-talc within the composite structure [36]. Furthermore, the degree of dispersion and the depth of intercalation are closely tied to the interactions between the polymer and nanoclay [36] and are influenced by the chosen preparation method [37]. Consequently, the XRD data aligns with and substantiates the findings observed through TEM analysis.

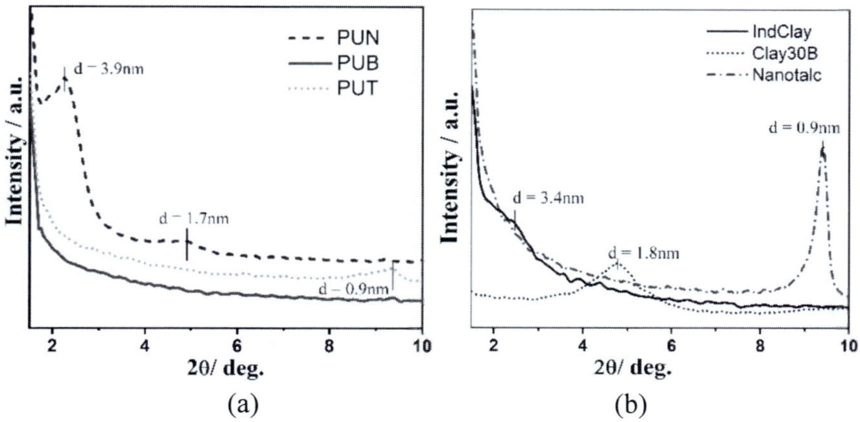

Figure 4. X-Ray diffraction of PU nanocomposites incorporating talc (PUT), Indian clay NK75 (PUN) and C30B (PUB) as fillers; (b) XRD of the three fillers: nanotalc, Indian clay NK75 (Indclay) and Cloisite 30B (clay 30B). Note: Figures are reproduced using data from [14] and [26].

For the polymer nanocomposites incorporating Indian clay (NK75), the peak at $2\theta = 2.4^{o}$ (corresponding to a d-spacing of 3.6 nm) observed in pure Indian clay is intensified in the nanocomposites, with no alteration in the 2θ value and d-spacing. This increase in d-spacing is ascribed to the intercalation of silicate layers and polymer chains, while the heightened intensity is likely a result of the stacking of silicate layers within the nanocomposites. These observations provide valuable insights into the structural changes and interactions occurring in the nanocomposite materials.

In Figure 5a, the FTIR spectra of pure TPU, PUB, PUT, and PUN are presented. The –NH and –OH peak appears at 3327 cm^{-1}, attributable to the hydrogen-bonded –NH in the urethane linkage, while carbonyl –C=O stretching is observed at 1725 cm^{-1}, indicating hydrogen-bonded carbonyls [38]. The –NH peak of pure PU is situated at 3333 cm^{-1}, and –C=O peaks are at 1725 cm^{-1}. Typically, carbonyl peaks around 1700 cm^{-1} are challenging to resolve due to the superposition of peaks from urethanes and ester-polyols. Here, two peaks are considered, arising from the carbonyls of urethanes and ester-polyols in pure PUs. When carbonyl is bonded to hydrogen, it has a peak at 1703 cm^{-1}, and when it is free, it has a peak at 1728 cm^{-1} [14]. In the FTIR spectra of nanocomposites, a singular stretching is evident, accompanied by a significant reduction in the hydrogen-bonded peak. This reduction is because of the interactions between the filler particles and the PU matrix. For PU, two peaks are observed at 3332 cm^{-1} and 3304 cm^{-1}, representing non-hydrogen-bonded and hydrogen-bonded –NH peaks, respectively. In PUB, the hydrogen-bonded peak is more prominent, while the non-hydrogen-bonded peak is nearly invisible, indicating extensive hydrogen bonding between the –NH of urethane and the –OH present in Cloisite 30B. In PUT nanocomposites, the hydrogen-bonded and non-hydrogen peaks are identical to those in PU. Both the suppression and shifting of peaks indicate interactions among the components of the nanocomposites. The lack of a non-hydrogen-bonded peak in PUB implies that these nanocomposites exhibit superior interaction among their components compared to PUT nanocomposites. For PUN, the non-hydrogen-bonded peak is more pronounced than the hydrogen-bonded peak, with a shift from 3332 cm^{-1} in PU to 3326 cm^{-1} in PUN. This shift in the –NH peak is attributed to the dipole-dipole interaction between PU and nanoclay.

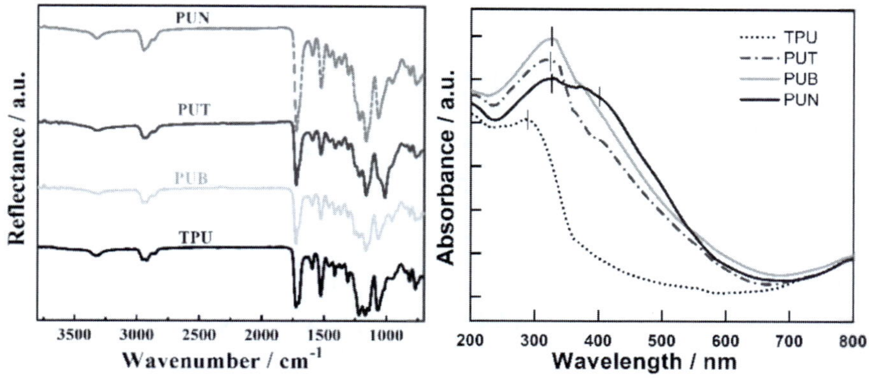

(a) (b)

Figure 5. The figure (a) shows the FTIR spectra and (b) UV-visible spectra of TPU and its nanocomposites prepared using Talc (PUT), Indian clay NK75 (PUN) and C30B (PUB). (Reproduced using data and figures from [14] and [26]).

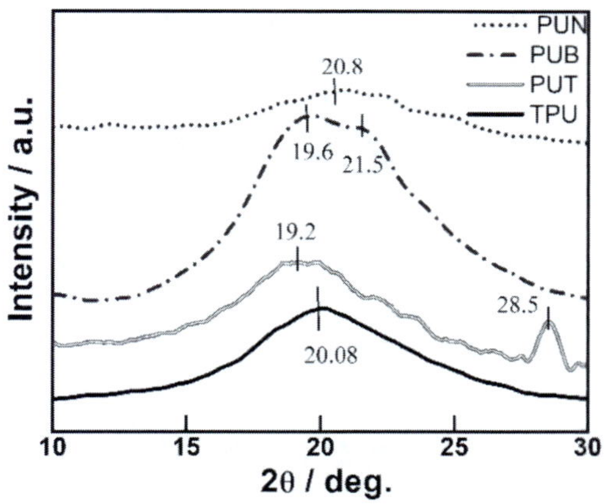

Figure 6. X-ray diffraction of thermoplastic polyurethane along with its PU-nanotalc, PU-NK75 and PU-Cloisite 30B nanocomposites. (Reproduced using the data and figures from [14] and [26]).

5.3.3. UV-Vis Spectroscopy

Figure 5b presents the UV-visible spectra of the nanocomposites. Polyurethane (PU) has a large hump peak visible in its UV spectrum at 288 nm [14]. This is explained by the molecular makeup of polyurethane, which has di-isocyanate groups with a double bond between carbon and oxygen (C=O). An absorption band is created via the $\pi - \pi^*$ electronic transition, and it appears at about 300 nm [39].

Upon the incorporation of the filler, a noticeable red shift in the peaks is observed, accompanied by increased absorption. The maximum red shift is observed in PU-NK75 (PUN) nanocomposites. The peaks in PU-Talc (PUT) nanocomposites manifest around 319 nm, while in PU-C30B (PUB) nanocomposites, the peak is seen at 326 nm. Additionally, in PUT nanocomposites a distinct peak emerges approximately at 400 nm, likely attributed to the presence of the filler.

This spectral analysis provides insights into the changes induced by the filler in the UV-visible characteristics of the nanocomposites, offering a deeper understanding of the molecular interactions and structural alterations in the composite materials.

5.3.4. XRD

Figure 6 shows that the pristine PU polymer exhibits a broad hump [14], likely indicative of the amorphous nature of the polymer [1]. Upon the incorporation of the filler particles, a crystalline peak emerges in all the polymer nanocomposites at $2\theta = 21.5^{\circ}$. The distinct peaks in XRD pattern indicate that the filler particles induce a certain level of crystallinity in the polymer matrix. Additionally, as shown in Figure 6, the peaks at $2\theta = 28.5^{\circ}$ in PU-Talc nanocomposites correspond to the nanotalc peak, resembling the peaks found in the XRD data of the pure fillers.

6. Mechanical Properties

The tensile test, also known as the tension test [11], stands as the most fundamental mechanical test, offering valuable insights into material behavior. This test is not only relatively inexpensive but also simple and fully standardized. It plays a pivotal role in determining how a material responds to tensile forces, unraveling crucial information about its strength and elongation characteristics as it undergoes tension. The maximum elongation of the specimen and the ultimate tensile strength are important characteristics that are directly measured by the tensile test. Additionally, Young's modulus, strain-hardening characteristics, yield strength, and Poisson's ratio can be derived from these measurements.

Tensile testing machines fall into two main categories: hydraulic and electromechanical. The electromechanical machine utilizes an assembly of gear reduction system, electric motor, and screws (one, two, or four in number) to impart up- and-down motion to the crosshead. The speed of the crosshead can be precisely controlled using a closed-loop servo system with a range of crosshead speed options. The hydraulic testing machines, on the other hand, employ a single or dual-acting piston to execute the upward and downward movement of the crosshead. In manually operated machines, a needle valve is used by the operator to control the rate of loading of the specimen. In a closed-loop hydraulic servo system, an electrically operated servo-valve ensures precise control, replacing the traditional needle valve. While the electromechanical machine offers versatility with a wide range of long crosshead displacements and test speeds, the hydraulic machine offers cost-effective generation of high forces by adjusting the speed of the motor. Both classes of testing machines contribute to the comprehensive understanding of material behavior under tension.

6.1. Specimen Preparation

6.1.1. Injection Molding

Injection molding is generally used for tensile test specimen preparation. It stands as one of the most prevalent manufacturing processes for plastic components, accommodating both thermoset and thermoplastic polymers. The process initiates by taking polymer granules or pellets and heating them to a molten state. Subsequently, the molten material is injected into a chamber created by a split-die mold. The molten substance remains in the mold, undergoing either heating for thermoset curing or cooling for thermoplastic solidification. Following this, the mold is opened, and the produced part is extracted. Injection molding offers advantages such as high throughput, low operational costs, and the flexibility to create components with intricate shapes. Despite its relatively expensive tooling cost, injection molding remains one of the most favored manufacturing processes for plastic materials.

A standardized sample cross-section, known as a tensile specimen or dog-bone specimen features two broad shoulders with a gage section situated between them is used for tensile testing. The shoulders are designed to provide a secure grip, while the gage section, with a smaller cross-section, allows for deformation and eventual failure. The manufacturing of shoulders for the dog bone specimen can be executed in various ways to accommodate different grips in the tensile testing machine, each with its own set of advantages and disadvantages.

For instance, serrated grips offer alignment dependency on the technician's skill, but they are easy and cost-effective to manufacture. But the pinned grips ensure excellent alignment, and threaded shoulders and grips also guarantee good alignment. However, in the case of threaded shoulders, technicians must thread each shoulder into the grip at least one diameter's length to prevent thread stripping before fractures occur.

6.2. Results

Tensile tests were conducted using a universal testing machine (Instron 5000). The specimens, shaped like a dog bone, had a gage length of 20 mm, thickness of 2.12 mm, and width of 4 mm, produced through the injection molding technique. The strain rate employed was 5 mm/min. Young's modulus was calculated employing Hooke's law, i.e., from the slope of the stress-strain curve up to the elastic limit., and the area under the curve represented the toughness.

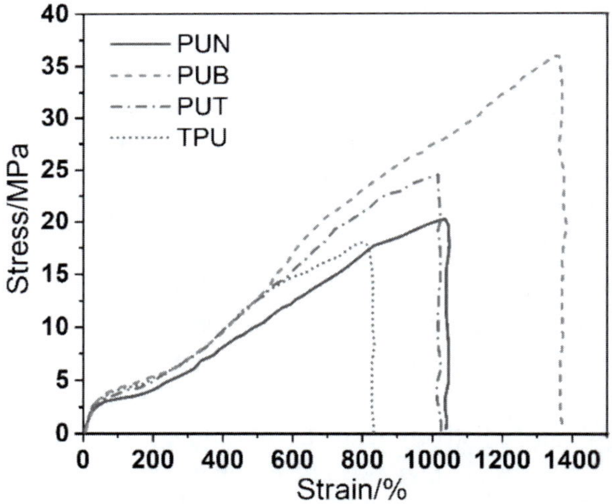

Figure 7. The Stress-strain curves for polyurethane and its PU-nanotalc (PUT), PU-NK75 (PUN) and PU-Cloisite 30B (PUB) nanocomposites. (Reproduced using data and figures from [14] and [26]).

The stress-strain curves obtained from tensile tests conducted on PU and PU/filler nanocomposites with a filler concentration of 4% by weight are presented in Figure 7. Notably, PUN nanocomposites exhibit the most substantial improvement in Elastic Modulus. The tensile modulus of the nanocomposites demonstrates an increase with the rising filler concentration, attributed to the dipole-dipole interaction between the matrix and the filler; see Figure 8. The highest enhancement in the mechanical behavior of PUN nanocomposites subjected to tensile loading is ascribed to the stacking of silicate layers in nanocomposites and improved interaction between the matrix and filler.

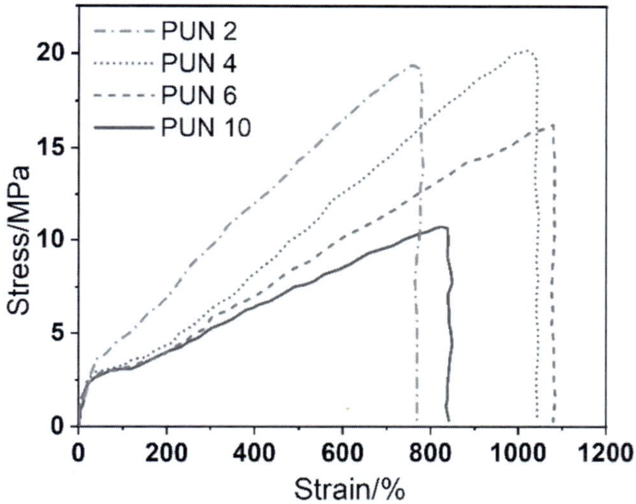

Figure 8. The figure shows the stress-strain curves for different %wt. of Indian clay NK75 for PU-NK75 nanocomposites.

In the mechanical property analysis of PUB nanocomposites at various filler concentrations, it is seen that the modulus of elasticity in tensile loading increases with the filler concentration, and elongation at break and toughness are higher for these nanocomposites, see Figure 9. This enhancement in mechanical properties is linked to the effective dispersion of silicate layers in the PU matrix and the interaction between the –OH group present in the filler and the –NH of the urethane. Examining the tensile test results of PUT nanocomposites (see Figure 10) with different filler concentrations reveals an enhancement in the modulus of elasticity with the rise in filler concentration, even in the absence or minimal interaction between the matrix and the filler. This modulus increase may be attributed to a slight enhancement in the crystalline nature as a result of the existence of nanotalc particles. The variation in modulus and toughness of the nanocomposites concerning the pure polymer is depicted in Figure 11.

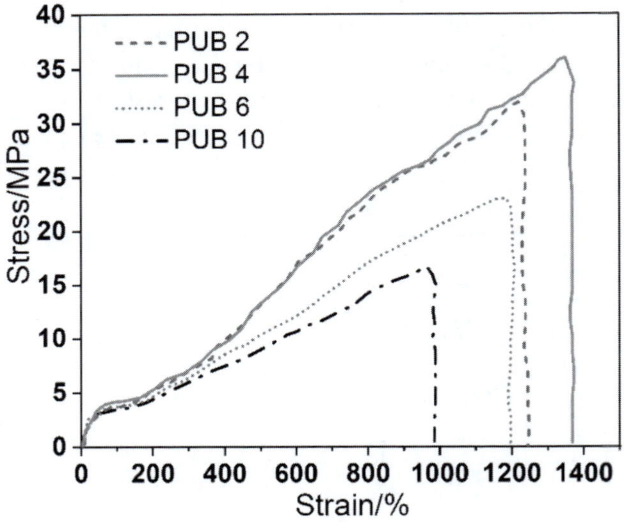

Figure 9. Stress-strain curves obtained for different %wt. of filler (C30B) for PU-Cloisite30B nanocomposites.

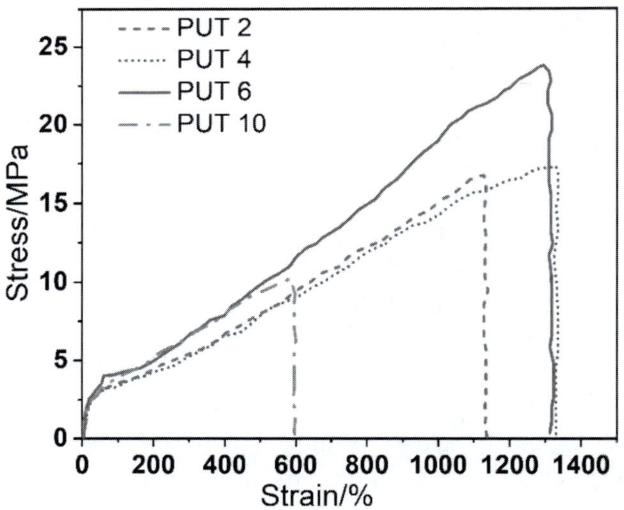

Figure 10. Stress-strain curves for PUT nanocomposites prepared incorporating different %wt. of nanotalc.

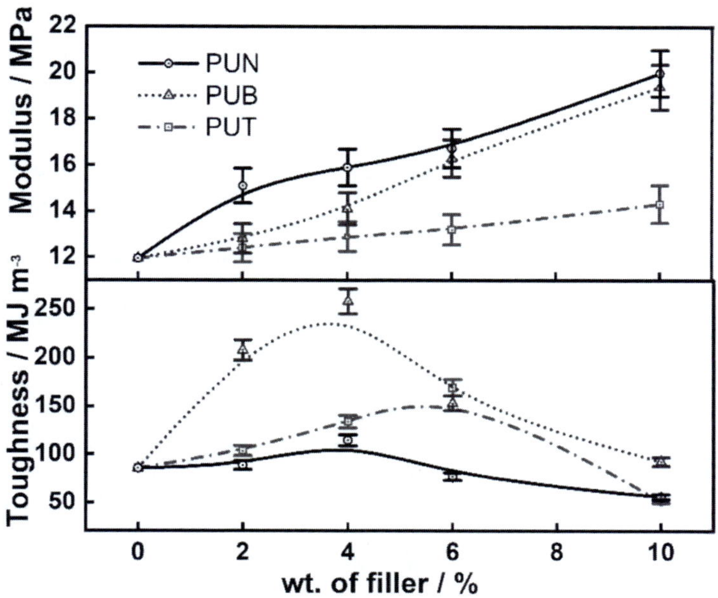

Figure 11. Modulus of elasticity and toughness obtained from stress-strain data of for PU-nanotalc, PU-Cloisite 30B and PU-NK75 nanocomposites incorporating different %wt. of filler. (Reproduced using data and figures from [14] and [26]).

Conclusion

The thermoplastic polyurethane (TPU) nanocomposites, fabricated through melt extrusion using diverse concentrations of talc and organically modified nanoclay as filler in the ester-type PU, outperformed the mechanical properties of pure TPU. The stiffness values exhibited a continuous increase with escalating nanoclay concentration, highlighting the superior mechanical response of the nanocomposites, accompanied by significant intercalation, driven by a robust interaction between the polymer matrix and the filler. This interaction was further substantiated through spectroscopic measurements, showcasing shifts in peak positions in both UV–vis absorption spectra and FTIR spectroscopy data.

Comparatively, PU-NK75 nanocomposites displayed superior mechanical responses when compared to PU-Cloisite 30B and PU-talc nanocomposites. The enhanced mechanical properties of these nanocomposites position them for various applications across different industries.

References

[1] Koerner, H., J. J. Kelley, R. A. Vaia, Transient microstructure of low hard segment thermoplastic polyurethane under uniaxial deformation, *Macromolecules* 41 (13) (2008) 4709–4716.

[2] Li, F., J. Hou, W. Zhu, X. Zhang, M. Xu, X. Luo, D. Ma, B. K. Kim, Crystallinity and morphology of segmented polyurethanes with different soft-segment length, *Journal of Applied Polymer Science* 62 (4) (1996) 631–638.

[3] Chattopadhyay, D. K., K. Raju, Structural engineering of polyurethane coatings for high performance applications, *Progress in polymer science* 32 (3) (2007) 352–418.

[4] Buckley, C., C. Prisacariu, C. Martin, Elasticity and inelasticity of thermoplastic polyurethane elastomers: Sensitivity to chemical and physical structure, *Polymer* 51 (14) (2010) 3213–3224.

[5] Randall, D., S. Lee, *The polyurethanes book* (2003).

[6] Mishra, A., B. P. Das Purkayastha, J. K. Roy, V. K. Aswal, P. Maiti, Nanoparticle controlled self-assembly in varying chain extended polyurethanes as potential nanobiomaterials, *The Journal of Physical Chemistry C* 116 (3) (2012) 2260–2270.

[7] Osman, M. A., V. Mittal, M. Morbidelli, U. W. Suter, Polyurethane adhesive nanocomposites as gas permeation barrier, *Macromolecules* 36 (26) (2003) 9851–9858.

[8] Choi, J. T., D. H. Kim, K. S. Ryu, H.-i. Lee, H. M. Jeong, C. M. Shin, J. H. Kim, B. K. Kim, Functionalized graphene sheet/polyurethane nanocomposites: effect of particle size on physical properties, *Macromolecular Research* 19 (8) (2011) 809–814.

[9] Mahanta, A. K., V. Mittal, N. Singh, D. Dash, S. Malik, M. Kumar, P. Maiti, Polyurethane-grafted chitosan as new biomaterials for controlled drug delivery, *Macromolecules* 48 (8) (2015) 2654–2666.

[10] Shah, D., P. Maiti, E. Gunn, D. F. Schmidt, D. D. Jiang, C. A. Batt, E. P. Giannelis, Dramatic enhancements in toughness of polyvinylidene fluoride nanocomposites via nanoclay-directed crystal structure and morphology, *Advanced Materials* 16 (14) (2004) 1173–1177.

[11] Jana, K. K., N. K. Vishwakarma, B. Ray, S. A. Khan, D. K. Avasthi, M. Misra, P. Maiti, Nanochannel conduction in piezoelectric polymeric membrane using swift heavy ions and nanoclay, *RSC advances* 3 (17) (2013) 6147–6159.

[12] Adak, B., B. S. Butola, M. Joshi, Effect of organoclay-type and clay- polyurethane interaction chemistry for tuning the morphology, gas barrier and mechanical properties of clay/polyurethane nanocomposites, *Applied clay science* 161 (2018) 343–353.

[13] Osman, M. A., V. Mittal, M. Morbidelli, U. W. Suter, Polyurethane adhesive nanocomposites as gas permeation barrier, *Macromolecules* 36 (26) (2003) 9851–9858.

[14] Pandey, S., K. K. Jana, V. K. Aswal, D. Rana, P. Maiti, Effect of nanoparticle on the mechanical and gas barrier properties of thermoplastic polyurethane, *Applied Clay Science* 146 (2017) 468–474.

[15] Zeng, Q., D. Wang, A. Yu, G. Lu, Synthesis of polymer–montmorillonite nanocomposites by in situ intercalative polymerization, *Nanotechnology* 13 (5) (2002) 549.

[16] Das, P., S. Manna, A. K. Behera, M. Shee, P. Basak, A. K. Sharma, Current synthesis and characterization techniques for clay-based polymer nano-composites and its biomedical applications: A review, *Environmental Research* 212 (2022) 113534.

[17] Silva, A. A., K. Dahmouche, B. G. Soares, Nanostructure and dynamic mechanical properties of silane-functionalized montmorillonite/epoxy nanocomposites, *Applied Clay Science* 54 (2) (2011) 151–158.

[18] Kumar, V., M. Kumar, G. Pugazhenthi, Effect of nanoclay content on the structural, thermal properties and thermal degradation kinetics of PMMA/organoclay nanocomposites, *International Journal of Nano and Biomaterials* 5 (1) (2014) 27–44.

[19] Alvim, R. S., C. R. Miranda, First principles characterization of silicate sites in clay surfaces, *Physical Chemistry Chemical Physics* 17 (7) (2015) 4952–4960.

[20] Ruiz-Hitzky, E., M. Darder, F. M. Fernandes, B. Wicklein, A. C. Alcântara, P. Aranda, Fibrous clays based bionanocomposites, *Progress in polymer science* 38 (10-11) (2013) 1392–1414.

[21] Norrish, K., The swelling of montmorillonite, *Discussions of the Faraday society* 18 (1954) 120–134.

[22] Kim, Y., J. L. White, Formation of polymer nanocomposites with various organoclays, Journal of applied polymer science 96 (5) (2005) 1888–1896.

[23] Nguyen, Q. T., D. G. Baird, Preparation of polymer–clay nanocomposites and their properties, Advances in Polymer Technology: *Journal of the Polymer Processing Institute* 25 (4) (2006) 270–285.

[24] Salahuddin, N., S. Abo-El-Enein, A. Selim, O. S. El-Dien, Synthesis and characterization of polyurethane/organo-montmorillonite nanocomposites, *Applied Clay Science* 47 (3-4) (2010) 242–248.

[25] Kapusetti, G., N. Misra, V. Singh, S. Srivastava, P. Roy, K. Dana, P. Maiti, Bone cement based nanohybrid as a super biomaterial for bone healing, *Journal of Materials Chemistry B 2* (25) (2014) 3984–3997.

[26] Pandey, S., P. Maiti, K. K. Jana, D. Rana, V. K. Aswal, Gas barrier properties of polyurethane nanocomposites, *Journal of Applied Polymer Science* 140 (32) (2023) e54256.

[27] Jamshidian, M., E. A. Tehrany, M. Imran, M. J. Akhtar, F. Cleymand, S. Desobry, Structural, mechanical and barrier properties of active pla–antioxidant films, *Journal of Food Engineering* 110 (3) (2012) 380–389.

[28] Najafi, N., M.-C. Heuzey, P. Carreau, Polylactide (pla)-clay nanocomposites prepared by melt compounding in the presence of a chain extender, *Composites Science and Technology* 72 (5) (2012) 608–615.

[29] Song, W., Z. Zheng, W. Tang, X. Wang, A facile approach to covalently functionalized carbon nanotubes with biocompatible polymer, *Polymer* 48 (13) (2007) 3658–3663.

[30] Tarapow, J., C. R. Bernal, V. A. Alvarez, Mechanical properties of polypropylene/clay nanocomposites: effect of clay content, polymer/clay

compatibility, and processing conditions, *Journal of applied polymer science* 111 (2) (2009) 768–778.

[31] Camargo, P. H. C., K. G. Satyanarayana, F. Wypych, Nanocomposites: synthesis, structure, properties and new application opportunities, *Materials Research* 12 (2009) 1–39.

[32] Finnigan, B., D. Martin, P. Halley, R. Truss, K. Campbell, Morphology and properties of thermoplastic polyurethane composites incorporating hydrophobic layered silicates, *Journal of applied polymer science* 97 (1) (2005) 300–309.

[33] Adak, B., M. Joshi, B. S. Butola, Polyurethane/clay nanocomposites with im- proved helium gas barrier and mechanical properties: Direct versus master- batch melt mixing route, *Journal of Applied Polymer Science* 135 (27) (2018) 46422.

[34] Mittal, V., Polymer layered silicate nanocomposites: a review, *Materials* 2 (3) (2009) 992–1057.

[35] Raquez, J.-M., Y. Habibi, M. Murariu, P. Dubois, Polylactide (pla)-clay nanocomposites, *Progress in Polymer Science* 38 (10-11) (2013) 1504–1542.

[36] Vaia, R. A., E. P. Giannelis, Polymer melt intercalation in organically-modified layered silicates: model predictions and experiment, *Macromolecules* 30 (25) (1997) 8000–8009.

[37] Wang, D., J. Zhu, Q. Yao, C. A. Wilkie, A comparison of various methods for the preparation of polystyrene and poly(methyl methacrylate) clay nanocomposites, *Chemistry of Materials* 14 (9) (2002) 3837–3843.

[38] Pattanayak, A., S. C. Jana, Thermoplastic polyurethane nanocomposites of reactive silicate clays: effects of soft segments on properties, *Polymer* 46 (14) (2005) 5183–5193.

[39] Wang, Y., H. Wang, X. Li, D. Liu, Y. Jiang, Z. Sun, O_3/UV synergistic aging of polyester polyurethane film modified by composite UV absorber, *Journal of Nanomaterials* 2013 (2013).

Chapter 5

Tensile, Compressive, Interlaminar Shear Strength Behavior of Lignocellulosic Fiber Reinforced Thermoplastic Polyester Matrix Composites

Naruemon Sumrith*

Department of Agricultural Engineering for Industry, Faculty of Industrial Technology and Management, King Mongkut's University of Technology North Bangkok (KMUTNB) Prachinburi Campus, Prachinburi, Thailand

Abstract

Polyesters are classified into thermosetting and thermoplastic types. Thermosetting polyesters, like alkyds and unsaturated polyesters, are used in coatings and formed with unsaturated components such as maleic anhydride. Thermoplastic polyesters, including PBT and PET, are created by combining aromatic dicarboxylic acids with aliphatic diols, used in fibers, films, and thermoformed parts. Natural fibers are beneficial for automotive applications due to their low density, renewability, and potential to reduce fuel consumption. Thermoplastic polyester composites are versatile materials widely utilized in automotive, electronics, and recreation industries. Their growth, marked by a 4-7% annual increase, is driven by the demand for lighter materials and metal replacements. Polyesters are divided into aliphatic, semi-aromatic, and aromatic subgroups, each with unique properties. Aliphatic polyesters, such as PLA, are notable for their biodegradability and use in medical applications. Semi-aromatic polyesters, like PET and PBT, are crucial in engineering and packaging due to their strength and stability.

---

* Corresponding Author's Email: naruemon.s@itm.kmutnb.ac.th

In: Mechanical and Thermal Properties of Engineering …
Editors: Senthilkumar Krishnasamy, Thitinun Ungtrakul, Sabarish Radoor et al.
ISBN: 979-8-89530-460-0

Aromatic polyesters are recognized for high-performance applications due to their rigidity and heat resistance. Lignocellulosic fibers, derived from renewable materials like wood and paper, enhance polymer composites. These fibers consist of cellulose, hemicellulose, and lignin, each contributing distinct properties. Cellulose provides structural strength, hemicellulose adds flexibility, and lignin offers protective functions. Tensile properties are crucial for understanding material responses to stretching forces, influencing tensile strength and elasticity. For fiber-reinforced polymer composites, the fiber-matrix interface significantly affects performance. Differences in properties, such as in PET-sisal composites, highlight the importance of good interfacial adhesion. Studies of natural fiber composites, such as sisal and jute, show reinforcement improves tensile strength and modulus. However, factors like humidity can impact performance, emphasizing careful material selection and processing. The compressive behavior of cellular solids reinforced with natural fibers shows that thickness and fiber type significantly impact performance. Honeycomb structures, particularly those reinforced with flax fibers, demonstrated superior compressive strength and lower strain. Incorporating sisal fibers in polylactic acid composites enhanced compressive strength, with treated fibers showing the highest improvements, highlighting the importance of fiber treatment and thickness in optimizing material properties. The study underscores the importance of optimizing interlaminar shear strength in lignocellulosic-thermoplastic composites by focusing on fiber-matrix interactions. Enhanced interphase properties contribute to better stress transfer, improving performance in various aspects, though increased interfacial shear strength may reduce overall toughness.

Keywords: thermoplastic polyester, lignocellulosic fiber, tensile properties, interlaminar shear strength

1. Introduction

Polyesters can be classified into two main types: thermosetting and thermoplastic. Within the category of thermosetting polyester, there are two distinct groups. The first group comprises alkyds, which are produced by condensing dibasic acids or anhydrides with polyhydric alcohols. Traditionally, the use of these polyesters in the coatings industry has grown to a large extent, particularly in the manufacturing of paints, lacquers, and varnishes. The second group consists of linear unsaturated polyesters, which are formed using unsaturated building blocks like maleic anhydride. Their molecular weights, typically falling within the range of 1,000 to 5,000, become apparent after a condensation process involving a diol. Consequently, their viscosity decreases due to dilution with an unsaturated monomer such as styrene or diallyl phthalate [1]. Thermoplastic polyesters are the result of combining aromatic dicarboxylic acids with aliphatic diols. Two primary products falling into this category are polybutylene terephthalate, often abbreviated as PBT, and polyethylene terephthalate, commonly denoted as PET or PETP. These polyester resin products have been subject to modification by introducing other diacids, including cycloaliphatic diacids, or diols. They have also been blended with other types of resins. Initially, polyester resins found use primarily as fibers and films, but over time, they evolved for applications as structural materials or shapes. One of their significant contemporary applications involves the production of thermoformed parts. While polyesters are recognized for their excellent dimensional stability at high temperatures, they do not exhibit particularly favorable wear and friction characteristics when compared to other thermoplastics [2].

A composite material is a term used to describe the combination of two or more substances with different chemical compositions, resulting in a final product possessing characteristics that surpass those of the individual constituent materials. Typically, the primary components of a composite material are the matrix and the reinforcing component, often referred to as the filler or reinforcement. The general properties of each type of material encompass both advantages and disadvantages. Consequently, various materials are blended to create products with unique properties derived from the strengths of each constituent material [3, 4].

Various reinforcing components can be integrated with the thermoplastic polyester (TPS) base to improve its physical, mechanical, and thermal characteristics. Assessments of TPS composite mechanical performance generally comprise evaluations of tensile strength, flexural strength, impact resistance, stiffness, hardness, and dynamic mechanical features, among other factors. Furthermore, the influence of reinforcement filler on essential thermal properties, e.g., temperature stability, heat conductivity, glass transition temperature (T_g), and melting temperature (T_m), has been thoroughly examined [5].

Many researchers and manufacturers have widely investigated and applied TPS composites, which are fabricated by incorporating both natural and synthetic materials as a reinforcing component within the TPS matrix. Reinforcement materials can be categorized into four fundamental groups:

fibers, particles, fillers, and flakes. Flakes, which are flat and two-dimensional in shape, can be effectively incorporated into glass or plastic to create composite materials. They are typically aligned in a parallel manner, resulting in higher density compared to fiber-based composites. Silver, aluminium, and mica are also common materials in the flakes. The mica flakes are easily machinable on a glassy matrix and can be used for electrical purposes. In the paint industry, aluminium flakes are used on a regular basis where they align parallel to the surface, such as in paints. Silver flakes are preferred for applications requiring high conductivity. Fillers are particles or powders integrated into materials that have enhanced the physical and mechanical characteristics of composite materials. They serve to reduce the use of more expensive bond materials. Fillers are usually used to alter or improve the properties of heat sensitivity, electrical resistance, friction, fatigue, and flame resistance e.g., aluminum oxides, calcium carbonate, limestone (calcium oxides), fluoride silicon, treated clay and hollow glass beads. The different forms of particulate matter used in composites include small particles (< 0.25 μm), carbon nanotubes, hollow spheres, cubes, and platelets. These particulates contribute favorable material characterization, while the matrix is used as a mandatory reinforced medium for construction purposes. Random or preferred orientation may be displayed in the configuration of these particulate materials. Particles generally do not have a high degree of effectiveness in increasing the strength or resistance to fractures. Chromium, lead, tungsten, copper, and molybdenum are typical particles used in composite materials. A fiber refers to a cord or thread incorporated into composite materials, typically characterized by a high aspect ratio (length/diameter), often exceeding 100. The fiber's cross-section can take on various shapes, including circular, square, or hexagonal. Fibers are divided into 2 different groups: synthetic fiber and natural fiber [6, 7]. Synthetic fibers offer advantages such as exceptional strength, dispersibility, improved ductility, and enhanced durability. Various types of synthetic fibers are extensively employed in industries such as aerospace and automotive. Over the last two decades, synthetic fibers such as carbon fibers, glass fibers, polyester fibers, aramid fibers, polyacrylonitrile fibers, polyamide fibers, and boron fibers have gained increasing attention [7, 8, 9].

Substituting glass fibers with natural fibers offers an advantageous option for the automotive industry. This is attributed to their low density and larger fiber diameter, which can enhance tensile strength with minimal filler content, resulting in lightweight materials. In essence, this can lead to reduced fuel consumption and is particularly beneficial for high-mileage vehicles.

Furthermore, the biodegradable, renewable, non-toxic, and abundant nature of natural fibers positions them as excellent candidates for future use in automotive composites [10].

2. Thermoplastic Polyester

A highly flexible category of composite materials is thermoplastic polyester composites that find widespread utility across a diverse spectrum of applications, encompassing but not limited to the automotive, appliance, cookware, electronics, and recreational sectors. The persistent growth in this field is underscored by annualized growth rates averaging 4-7%. This growth trajectory can be attributed to several driving factors, most notably the imperatives of weight reduction, component consolidation, and the substitution of traditional metallic materials with their thermoplastic counterparts. Polyesters, constituting a prominent and pivotal category within the domain of synthetic polymers, have garnered significant scholarly and industrial attention, largely owing to their exceptional adaptability to a wide array of applications. Within this overarching classification, polyesters may be systematically categorized into three primary subgroups: Aliphatic, Semi-aromatic, and Aromatic, each characterized by distinct structural and performance attributes.

Significantly, Aliphatic polyesters, constituting the foremost subgroup, have assumed a preeminent role, particularly within drug delivery applications. This eminence arises from their intrinsic qualities of biodegradability and biocompatibility. They serve as exemplars of synthetic biodegradable polymers, with their synthesis primarily predicated on the polycondensation of diols and dicarboxylic acids. Nevertheless, it is imperative to acknowledge the initial constraints that impeded their widespread applicability. These limitations encompass issues such as restricted hydrolytic stability, low melting points, and relatively modest molecular weights, collectively circumscribing their functional utility. Intriguingly, the heightened susceptibility of Aliphatic polyesters to hydrolytic degradation, while initially perceived as a limitation, has rendered them immensely valuable in the sphere of biomedical applications. Within this category, one encounters a range of notable polymer exemplars, including, but not confined to, polyglycolide (PGA), polylactic acid (PLA), polycaprolactone (PCL), polyhydroxyalkanoate (PHA), polyhydroxybutyrate (PHB), polyethylene adipate (PEA), polybutylene succinate (PBS), and poly(3-hydroxybutyrate-co-3-hydroxyvalerate) (PHBV). These polymers, through their distinctive attributes and biocompatibility, have advanced the frontiers of biomedicine and therapeutics [11, 12, 13]. Semi-aromatic polymers, characterized by ester groups in their main chain, have become an integral component of modern society's development and constitute the largest output share among synthetic polymers. The most commonly employed thermoplastic polyester resins in this category include poly(ethylene terephthalate) (PET), poly(butylene terephthalate) (PBT), and poly(1,4-cyclohexylenedimethylene terephthalate) (PCT). The recently added product to the market is poly (trimethylene terephthalate) (PTT), which has gained commercial traction because its monomer 1,3-propanediol can be readily available in massive quantities. These polyester materials, with the exception of metals and additional resin such as polycarbonate, poly (phenylene sulfide), polypropylene, and nylon [14, 15]. All-aromatic polyesters, commonly referred to as poly(arylates), have attracted substantial attention in both academic research and commercial domains due to their potential as fundamental constituents for high-performance polymeric materials. The nomenclature "poly(arylate)" conventionally designates polyester compounds characterized by the exclusive presence of all-aromatic moieties, with a particular emphasis on those composed of a balanced combination of TPA (Terephthalic Acid), IPA (Isophthalic Acid), and bisphenol A. These materials exhibit notably elevated glass transition temperatures, approximately hovering around 180°C. Two main categories can be systematically defined for all-aromatic or highly aromatic polyesters: firstly, the "amorphous high T_g materials," typified by the TPA/IPA bisphenol A polymer as described above; and secondly, the class of "liquid crystalline aromatic polyesters," in which structural properties are derived from the inherent rigidity and spatial packing capabilities found within the chains of aromatic polyesters [11, 16]. The first class of "Amorphous High Tg Materials," represented by the TPA IPA bisphenol A polymer described above, where structural properties are derived from its inherent rigidity and spatial packing ability within a chain of aromatic polyesters.

2.1. Polylactic Acid (PLA)

Polylactic acid (PLA) represents an aliphatic polyester characterized by the absence of an aromatic structure, rendering it inherently hydrophobic, although it exhibits a greater hydrophilicity in comparison to Polyethylene

Terephthalate (PET). PLA stands as a biodegradable thermoplastic polymer sourced from sustainable raw materials, notably corn starch and sugar cane. Its versatile composition encompasses several variants, including Poly-L-lactic acid (PLLA), poly-D-lactic acid (PDLA), and poly-DL-lactic acid (PDLLA), each tailored to specific applications such as the fabrication of pins, screws, plates, and rods. PLA is characterized by its semi-crystalline nature, notable for a relatively gradual rate of crystallization. The material has garnered particular favor within the realm of biomedical applications, primarily due to its pronounced biocompatibility, facile manufacturability, hydrophobic tendencies, and intrinsic biodegradability. A rising trend in the utilization of PLA extends to textile applications, where it offers enhanced comfort and superior physiological performance. Consequently, PLA finds application in sportswear and shirts, endowing wearers with an elevated sense of comfort and improved physical well-being, even under conditions of heightened physical activity [11, 17].

2.2. Polyhydroxyalkanoates (PHA)

Polyhydroxyalkanoates (PHA) constitute a class of polyesters that are enzymatically synthesized by a diverse array of bacteria, encompassing both Gram-positive and Gram-negative species. Remarkably, more than 75 distinct bacterial strains have been identified as proficient producers of PHA. These biopolymers serve as crucial intracellular reservoirs of both carbon and energy, a feature particularly pronounced under conditions of nutrient scarcity. This indicates that during these periods of nutritional stress, PHAs may be accumulated in bacterial cells to levels which are as high as 90% of the cell's dry weight. The controlled biosynthesis of PHAs in many bacterial strains can be meticulously regulated through the continuous provision of precursor molecules in the context of chemostat cultivation. This strategy is complemented by the imposition of multiple nutrient limitations, ensuring that no deleterious concentrations of substrates accumulate within the culture supernatant. Consequently, the bacterium's metabolic resources are exclusively channeled towards the production of PHAs, without being diverted for biomass formation. The ecological soundness and desirable material properties of PHAs have rendered them subjects of substantial interest within the realm of industrial applications, primarily as a sustainable alternative to traditional non-biodegradable plastics. Nevertheless, it is important to acknowledge that the current industrial utilization of PHAs is constrained by the relatively elevated production costs and the expenditures associated with downstream processing. Nonetheless, it is reasonable to anticipate that ongoing research and development endeavors aimed at optimizing large-scale production and cost-effectiveness will facilitate the expanded and more pervasive integration of PHAs within the industrial landscape [18, 19, 20].

2.3. Poly (Butylene Succinate) (PBS)

Poly(butylene succinate) (PBS) is available in the market, and aliphatic polyester with numerous interesting properties such as biodegradability, meltableness, heat and chemical resistance. The PBS, which is formed by polycondensation between sulfonic acid and butanediol, is one of the most important biodegradable polymers. There are two steps in the reaction. A first esterification occurs between the diacid and the diol, followed by a polycondensation at high temperature in order to produce higher molecular weight PBS. Depending on the monomers used, PBS has been entirely produced from biobased succinic acid. These products are used for consumer goods applications, electronics and foodstuffs packaging materials such as bowls, cups or plates. Due to its easy compostable nature, the use of PBS in agricultural applications has also begun [21, 22, 23].

2.4. Polyethylene Terephthalate (PET) 30/10/2023

Polyethylene terephthalate (PET) represents about 18% of the global plastic market and has been widely employed, with a number of applications. Around 50 million tonnes are produced annually. It is utilized in various industries, including soft drink bottles, packaging, electronic component, photovoltaic panel, textile, and more, owing to its unique properties such as a glass transition temperature (T_g) around 67°C, a melting temperature (T_m) close to 265°C, tensile strength of about 53 MPa, a tensile elongation of around 300%, and excellent barrier characteristics. PET is resistant to microbial attack and does not degrade under normal environmental conditions, necessitating rigorous chemical degradation methods. Recycling is promoted, even though full recovery remains challenging. PET offers advantages like lower raw material costs, higher strength, and temperature resistance. Proper processing and adherence to drying guidelines are crucial to prevent hydrolytic degradation during molding [14, 24, 25].

2.5. Poly(Butylene Terephthalate) (PBT)

Polybutylene terephthalate (PBT), an extensively utilized thermoplastic polyester, has garnered significant attention in the realms of engineering and biomedicine. Its production involves a chemical reaction between dimethyl terephthalate and 1,4-butanediol, typically employing organ titanate catalysts. Originally introduced as an automotive substitute for phenolic resins, PBT is characterized as a semicrystalline thermoplastic with a melting point of approximately 230°C, a feature subject to variation based on sample preparation and annealing conditions.

Distinguishing itself from its counterpart, polyethylene terephthalate (PET), PBT exhibits a notably accelerated crystallization rate. This property renders it particularly suited for specific applications where expeditious crystallization is advantageous. The glass transition temperature (T_g) of PBT spans a range of 30 to 50°C, with the potential to reach as low as 20°C. The precise T_g value is contingent upon factors such as crystallinity and the methodology employed for measurement.

The commendable attributes of PBT encompass its rapid crystallization kinetics, ease of processing, and outstanding mechanical properties. These characteristics have led to PBT's preferential use in the realm of injection molding and within fiber-reinforced semicrystalline composites. In many applications, PBT outperforms PET, particularly due to its exceptional mechanical strength, durability, high resistance to heat-induced distortion (with values reaching up to 215°C for glass fiber-reinforced PBT), swift crystallization rates, sustained usage at elevated temperatures, dimensional stability, and notably abbreviated injection molding cycle times.

Evidently, the global demand for PBT continues to ascend, substantiating its significance in contemporary industries. It is imperative to acknowledge that the properties of PBT are intrinsically linked to its crystallinity and the resulting morphological attributes following the processing phase [26, 27].

2.6. Poly(1,4-Cyclohexylenedimethylene Terephthalate) (PCT)

Poly(1,4-cyclohexylenedimethylene terephthalate) (PCT) emerges as a semicrystalline thermoplastic polyester characterized by a melting temperature encompassing the range of 251 to 318°C. The specific melting point within this range is contingent upon the trans/cis isomer ratio of its 1,4-cyclohexylenedimethyl (CHDM) constituents. PCT distinguishes itself with its notable attributes, including remarkable chemical resistance, thermal stability, impressive tensile strength, and fiber spinning capabilities that bear resemblance to poly(ethylene terephthalate) (PET). Consequently, PCT finds its niche within the realms of automotive and electronics/electrical applications, a testament to its exceptional suitability in these domains.

PCT's versatility is further underscored by its role as the foundational material for PCT-based copolyesters. These copolyesters exhibit a remarkable capacity for enduring high temperatures, and their properties can be fine-tuned through judicious compositional adjustments. Such adjustments are typically realized through the application of two prevalent methodologies: alcohol modification and acid modification. The former method, yielding alcohol-modified PCT copolyesters (PCTG), involves the partial replacement of CHDM with alternative diols. In the latter approach, commercial acid-modified PCT (PCTA) is produced by substituting a portion of the terephthalic units with isophthalic counterparts. It is imperative to note, however, that both commercially available PCT and its copolyesters share a common limitation—their flammability, which confines their utility within select applications [28].

2.7. Poly(Trimethylene Terephthalate) (PTT)

Poly(trimethylene terephthalate) (PTT) stands as a partially engineered thermoplastic polyester, manufactured via the condensation reaction of propane diol with terephthalic acid or dimethyl terephthalate. The glass transition temperature (T_g) of PTT spans a range of 40 to 75°C. Recent strides in technology have led to increased economic viability in the production of PTT, thereby affording new prospects across various industrial domains. These encompass but are not confined to the realms of textiles, carpeting, packaging, and a spectrum of engineering applications.

In the domain of engineering plastics, PTT emerges as a compelling contender, distinguished by a unique amalgamation of attributes. These encompass notable tensile strength, a capacity for elastic recovery, rapid crystallization kinetics, efficacious electrical insulation, enduring dimensional stability, commendable chemical resistance, and a remarkable ease of processing. These qualities underpin PTT's potential and its role within the pantheon of engineered materials [29, 30].

2.8. Reinforced Materials: Lignocellulosic Fibers

The careful choice of reinforcement materials within polymer composites and the robustness of the interfacial bond between the reinforcement and the polymer matrix assume pivotal significance in molding the physical, mechanical, and thermal attributes of TPS composites. Within the domain of TPS matrices, a diverse spectrum of materials is available to serve as reinforcements [31].

Lignocellulosic materials, sourced from natural and renewable reservoirs, assume a pivotal role in the operational framework of contemporary societies and hold profound significance in advancing a sustainable global economy. Notably, products derived from lignocellulosic origins, such as wood and paper, are fundamental drivers of societal progress. Nevertheless, the endeavor to enhance the quality of these products and refine manufacturing procedures encounters challenges primarily rooted in an insufficient comprehension of the intricate structures and chemical compositions inherent in these materials.

The intricate process of extracting cellulose molecules from lignocellulosic fibers unfolds through a succession of micro-scale operations, culminating in the transformation of fiber bundles into nanofibrils. The salient micro- and macrostructure characterizing plant cell walls, predominantly constituted of lignocellulosic components, assumes a paramount role in safeguarding plant survival. This multilayered cell wall construction is progressively established during the growth of plant cells, affording critical protection against an array of external stresses, encompassing mechanical, chemical, and microbial factors [32].

Lignocellulosic materials, as a rule, predominantly comprise cellulose (35–70%), hemicellulose (15–35%), and lignin (10–35%). It is imperative to recognize that the precise proportions of these components may fluctuate across distinct plant species. For example, heightened concentrations of lignin are typically encountered in more advanced plant taxa, whereas annual plants tend to exhibit diminished lignin content [33, 34, 35].

Cellulose, a linear homopolysaccharide, is composed of β-D-glucopyranose (glucose) units linked together through β-D-(1-4) bonds, with a chemical formula represented as $C_6H_{10}O_5$. This polysaccharide plays a pivotal role as a fundamental constituent of plant fibers, owing its notable hydrophilicity to the presence of hydroxyl groups (–OH) within its molecular structure. The ubiquity of cellulose in the natural world extends across both primitive and highly complex plant species. A common measure for the size of polymers or macromolecules is the degree of polymerization (DP, which characterizes the number of monomer units in a chain), and cellulose is distinguished by its typically substantial DP, often exceeding 10,000. Notably, in-depth conformational analysis has unveiled that cellobiose (4-O-β D-glucopyranosyl-β-D-glucopyranose) serves as the fundamental repeating unit within the cellulose polysaccharide, rather than glucose. Cellulose fibrils exhibit regions characterized by a highly structured, crystalline arrangement, alongside regions marked by a disordered, amorphous disposition. Nevertheless, the specific configuration and distribution of these crystalline and amorphous domains within cellulose fibrils continue to be subjects of active and ongoing research [34, 36, 37, 38].

Hemicellulose, which ranks among the most abundant materials alongside cellulose, is characterized by its complex and highly branched structure. It comprises various sugars, including glucose, glucuronic acid, mannose, arabinose, and xylose. A fundamental constituent of plant fiber materials, hemicellulose derives from plant polysaccharides and is characterized by a core chain comprising residues of D-xylose, D-mannose, D-glucose, or D-galactose. In addition to this central structure, hemicellulose exhibits branched chains formed by additional glycosyl units. These branched chains are intricately linked to the core structure, constituting a defining feature of hemicellulose. It is based on its various alkaline solubility compared to the cellulose that hemicellulose purification relies upon, which enables it to be extracted via an alkaline solution rather than without chloroform and pectin. Hemicellulose is a copolymer, composed of various quantities of different saccharide molecules, in contrast to cellulose. It boasts a degree of polymerization that typically falls within the range of 50 to 200, making it amorphous and susceptible to hydrolysis. Notably, when hemicellulose co-crystallizes with cellulose, it does not aggregate. There are variations between different plant sources in the content and structural composition of hemicellulose. The composition of the main and secondary chains of glucans in that polysaccharide is a major focus for research on hemicellulose's chemical structure. The primary chain may encompass one or multiple varieties of glycosyl units, each characterized by unique linkage patterns. The specific glycan compositions are contingent upon the plant's geographical origin and the specific plant parts used, necessitating a classification of glycans to facilitate an understanding of the chemical structure. Notable constituents of hemicellulose within the cell matrix include xylan, xyloglucan, glucomannan, mannan, galactomannan, callose, and others [34, 37, 38, 39].

In terms of the prevalence, lignin is second only to cellulose as one of the most widespread organic polymers in plant structures. It represents the exclusive chemical composition of both gymnosperms and angiosperms. Lignin content in wood typically falls within the range of 20% to 40%, while in Gramineae plants, it averages between 15% and 20%. The term "lignin" encompasses a group of substances, and variations are observed due to different plant species, growing seasons, and specific plant parts. In addition, lignin structures have a remarkably diverse character even in the presence of separate cell morphologies with identical xylem or other cells wall layers. Lignin presents itself as a highly intricate and amorphous structure, functioning as a binding material that occupies the spaces between cellulose and hemicelluloses. This polyphenolic macromolecule comprises three primary phenylpropane monomeric units: coniferyl alcohol (G), p-hydroxyphenyl alcohol (H), and sinapyl alcohol (S). Softwoods are characterized by a prevalence of the coniferyl alcohol structure, whereas hardwoods commonly feature a combination of sinapyl alcohol and coniferyl alcohol structures, with sinapyl alcohol dominating. In grasses, the structure of lignin primarily revolves around the p-hydroxyphenyl alcohol unit. This abundant phenolic polymer resides within plant cell walls, playing a pivotal role in facilitating water transport and providing essential protection to plants against various environmental stresses [34, 37, 38, 39, 40].

2.9. Tensile Properties

Tensile characteristics are essential parameters that reveal how a material reacts when exposed to stretching forces. These characteristics involve vital factors like the elasticity modulus, elastic threshold, elongation, proportional limit, reduction in size, tensile strength, yield point, and yield strength. It is crucial to explore tensile properties since they offer valuable understandings into how a material behaves when subjected to tensile stress. The evaluation of tensile properties adheres to established standards, often dictated by the American Society for Testing and Materials (ASTM). The specific ASTM standard utilized depends on the type of polymer composite under scrutiny. ASTM D638 is the recommended standard for composites characterized by random orientation, low reinforcement volume, or moldable attributes. Conversely, ASTM D3039 is the preferred choice for highly oriented and/or fiber-reinforced polymer composites with a high tensile modulus. In the realm of fiber-reinforced polymer (FRP) composites, an intriguing dichotomy exists between the constituent components. Typically, the fibers within FRP composites exhibit elevated strength and stiffness, while the matrix material demonstrates superior elongation capabilities. The amalgamation of these contrasting attributes culminates in the development of an FRP composite material with tensile properties that lie within the spectrum defined by the individual properties of its constituent elements [41].

The natural fibers used in composites exhibit inherent polarity, while PET is a relatively non-polar polymer. This polarity mismatch leads to poor adhesion at the fiber-polymer interface, with the fibers acting as stress concentrators in the composite. This issue is further exacerbated by the differences in failure strains between sisal fibers and the PET matrix. Neat PET displays a ductile behavior during tensile testing, with an elongation of 3.7%. In contrast, sisal-reinforced PET composites exhibit brittle behavior, fracturing at less than 0.5% strain, signifying that composite failure occurs before effective stress transfer from the fiber to the matrix. This emphasizes the critical role of the fiber-matrix interface in determining the mechanical properties of natural fiber-reinforced polymer composites. Conversely, the modulus of PET significantly increases with the addition of sisal fibers, especially as the volume fraction of sisal fiber increases. For instance, the tensile modulus of PET improves from 2.7 GPa to 6.4 GPa with the incorporation of 40% untreated sisal fiber and further increases to 7.9 GPa with alkali-treated sisal and 9.0 GPa with a combination of alkali and acetylated sisal. This enhancement in stiffness can be attributed to a robust interfacial interaction between the fibers and the polymer. The increase in tensile modulus is also influenced by the greater restraint on the mobility of the amorphous polymer region, as the fibers are stiffer than the polymer matrix [42].

The study focused on incorporating sisal and jute fibers into PLA composites. The results show that PLA laminate has higher tensile strength and rigidity compared to polypropylene (PP). The tensile strength of PP and PLA has been enhanced by the addition of these reinforcement fabrics. Initially, when composites acted as homogeneous materials with fibers and matrix being subjected to identical strain, the stress-strain curve showed a linear correlation in the elasticity range. In the context of jute/PP and sisal/PP composites, the transition from the initial elastic phase transpired at an approximate threshold of 10 megapascals (MPa). This transition was accompanied by a notable reduction in moduli, yielding values within the range of 2.7-2.9 gigapascals (GPa). Subsequently, the initiation of non-linear deformation became evident at approximately 33-38 MPa, culminating in the eventual structural failure of the composites at stress levels spanning from 45 to 52 MPa. The change for the PLA composite material was at about 20 MPa higher than that of the relevant PP composites in order to demonstrate improved adhesion between natural fibers and PLA. Compared to the PP composites, PLA produced a decrease in deformation which resulted in clear fractures. The tensile parameters have been adjusted to a 40% fiber volume fraction to take into account the variations in area density of reinforcing fabrics and laminates. Similar trends were observed in the tensile strength values of PLA composites with different reinforcement components compared to their respective PP composite counterparts. The normalized initial tensile moduli of jute, quantified at 9.0 GPa, and sisal composites, recorded at 9.9 GPa, exhibited a notable degree of comparability with the initial tensile moduli documented for glass composites, amounting to 9.8 GPa. Interfacial adhesion was slightly better in PLA composites, with lower fiber pull outs and a more uniform fracture. This is due to the differences in chemical structures where jute and sisal are cellulosic fiber with hydroxyl groups on their surface, which means that they have a low tolerance for nonpolar polymers such as PP. Conversely, polylactic acid (PLA) represents an aliphatic ester polymer characterized by the presence of oxygen atoms within its molecular structure. This inherent chemical composition renders PLA a more polar polymer, thereby affording superior interfacial adhesion to jute and sisal fibers [43].

The present study was centered on an in-depth exploration of the interplay between fiber loading and water absorption concerning the mechanical characteristics of a green composite. This composite entailed the integration of sugar palm fibers (SPF) into a matrix comprising sugar palm starch (SPS). Notably, the tensile strength of the composite exhibited a substantial and progressive increase, particularly when considering fiber loading up to 30%. This observed trend established a direct, positive correlation between the augmentation of fiber content and the subsequent rise in tensile strength. Specifically, tensile strength values were recorded at 2.42 MPa (for SPS), 3.12 MPa (for SPF10), 4.01 MPa (for SPF20), and 5.31 MPa (for SPF30). The remarkable enhancement in mechanical properties serves as a testament to the effectiveness of SPF as a robust reinforcing element within these composites. This remarkable performance can be ascribed to the efficient and precise transmission of stress between individual SPFs. Notably, this synergy reaches its pinnacle at a 30% fiber loading, signifying the optimized interaction between SPF and the surrounding matrix. However, it is prudent to highlight that the tensile strength of the sugar palm green composite displayed a noticeable decline after exposure to a high-humidity environment of 75% relative humidity over a duration of 72 hours. From this vantage point, it is reasonable to deduce that the SPF/SPS green composites eloquently demonstrate the favorable dispersion and adhesion of SPF within the SPS biopolymer matrix, albeit with an awareness of the composite's sensitivity to prolonged high-humidity conditions [44].

3. Compressive Behavior

The versatile attributes of cellular solids have sparked significant interest in their utilization within extremely lightweight structures. This investigation examines the compressive and failure characteristics of diverse honeycomb structures, encompassing multi-honeycomb, re-entrant honeycomb, and silicomb variants. These structures exhibit varying core heights of 10, 20, and 30 mm and are reinforced with a range of natural fibers, including flax, jute, and sisal. The moulds used to produce these composite core assemblies have been manufactured using 3D printing technology and were composed of a lactic acid PLA as the main material. In order to exhaustively assess the impact of several factors on final compression stress, deformation characteristics and energy absorption properties a series of experiments were carried out in accordance with Taguchi's Design L9. A range of 25.32 to 36.38 MPa was observed for the ultimate compression stress, while energy absorption values varied between 22.34 and 33.76 kj/m^3. Compressive strain ranged from -30.26 to -13.43 MPa, and the effect of the "fiber type and thickness" factors, as per Taguchi design, was examined. The data revealed variations in compressive strain across all structures, depending on fiber type and thickness. Notably, an increase in thickness (from 10 to 30 mm) led to a decrease in compressive strain for all structures, with honeycomb structures outperforming re-entrant and silicomb structures. Furthermore, flax fibers exhibited superior performance compared to jute and sisal. Finally, the most important parameter affecting compressive strain has been determined as thickness, and it was established that thickness significantly affects compression properties and failure behavior [45].

This research initiative examines the impact of an economical and environmentally friendly sisal fiber treatment prior to entering into lignocellulosic composition based on polylactic acid. The composition's characteristics were assessed concerning extractives, hemicellulose, lignin, and cellulose content. In order to remove sodium ions, the fibers were treated for different periods of 24, 72, 120 and 168 hours before washing them with distilled water. To address the growing demand for bio-composites, it is imperative to employ processes that offer rapid processability, precision, and repeatability. Extrusion injection molding is considered among these methods to be a commercially viable method for the production of bio-composites meeting those criteria. Hence, with a primary emphasis on the commercial viability of bio-composites, bio-composites comprising PLA and sisal were manufactured through the extrusion-injection molding process. A low sisal fiber concentration of 30% by weight has been chosen for this study, in order to follow up the results obtained from previous experiments. The interaction of the fibers and matrix influences, in particular, the compression strength of developed bio-composites. The presence of a strong connection between sisal fibers and PLA matrix results in increased compressive strength. In comparison to neat PLA, which exhibited a compressive strength of 81.4 MPa, all the developed bio-composites displayed improved compressive strength when sisal fibers were incorporated. Within this cohort, the PL/SF/72 bio-composites exhibited the most elevated compressive strength, attaining a value of 101.8 MPa. This stands in contrast to the compressive strength observed in untreated sisal fiber-reinforced PL/SF bio-composites, which measured 87.7 MPa [46].

4. Interlaminar Shear Strength Behavior

For the assessment of interface and interphase properties in order to optimize composite characteristics, it is important to take account of fibers interaction and adhesion with matrix on composite materials. This discussion delves into the current methodologies employed for the evaluation of these properties, with a specific focus on lignocellulosic-thermoplastic composites. Interactions between fibers and the matrix as well as interphase regions are essential in this composite for its ability to deliver stresses to fibers. This efficient stress transfer is particularly critical in shaping the properties of short-fiber composites. Furthermore, an augmentation in the interplay between fibers and the matrix leads to enhancements in various facets, encompassing off-axis performance, fatigue behavior, environmental durability, compression properties, and resistance to creep within the composite material.

Nevertheless, it is imperative to recognize that an elevation in interfacial shear strength (IFSS) may concomitantly precipitate a decline in the overall toughness of the composite material. Therefore, optimizing the interphase region's properties becomes a vital step in achieving the desired composite characteristics. In the context of the present study, hardwood birch dowels underwent a treatment process involving rinsing in distilled water, followed by drying at 80°C. A low molecular weight polyethyleneE43 copolymer and a low molecular weight ethylene-acrylic copolymer were used as matrix materials. Anionic emulsions of maleic anhydride-grafted polypropylene and maleic anhydride-grafted polyethylene were utilized to modify the fiber surface, serving as compatibilizers within cellulosic fiber-thermoplastic composites. Additionally, an emulsion of an ethylene and acrylic acid copolymer was used for fiber surface modification. The results revealed that dowels treated with E43 exhibited the highest interfacial shear strength when polyethylene was utilized as the matrix. Although the other treatments also yielded improvements, their IFSS values were comparatively lower than those of the E43-coated samples. The treatment incorporating the ethylene-acrylic copolymer yielded the most substantial shear strength enhancement. This can be chiefly ascribed to the intensified secondary interactions facilitated by the copolymer's polar constituent, acrylic acid, with the highly polar surface of the cellulosic material. Acid-base interactions were found to significantly enhance interphase properties. Consideration must be given to the fact that any changes in the surface characteristics of the dowel might affect the orientation of polymer molecules near the surface, which may possibly contribute to variations observed intermuscular shear strength [47].

Conclusion

The paper provides a comprehensive overview of composites that employ thermoplastic polyester matrices reinforced with lignocellulosic fibers. It underscores the significant advantages of hybrid composites, which are crafted by blending two or more varieties of lignocellulosic fibers within a polymer matrix. These composites manifest remarkable mechanical properties, indicative of robust interfacial adhesion between the lignocellulosic fibers and the polyester matrix.

Biopolymer green composites, when fortified with lignocellulosic fibers, exhibit outstanding mechanical performance, rendering them a promising candidate for replacing conventional composites. This is particularly relevant in secondary structural applications characterized by low to moderate impact requirements. To augment the mechanical attributes of green composites, hybridization with diverse lignocellulosic fibers within a biopolymer matrix is employed. The application of synergistic techniques to achieve an optimal fiber orientation and sequence in high-performance green composites results in appealing mechanical characteristics that can be tailored to specific needs.

Moreover, further advancements can be made through the hybridization of lignocellulosic fiber-reinforced green composites, a process involving the judicious selection of fibers and the application of suitable chemical treatments. It is noteworthy that composites combining kenaf, bamboo, and coir fibers with a polylactic acid matrix demonstrate the highest mechanical performance. Consequently, these hybrid green composites offer mechanical properties on par with synthetic composites, such as glass fiber composites. They come with the added advantages of eco-friendly manufacturing processes, cost-effective production, and affordable raw materials, positioning them as a viable alternative to traditional composite materials.

The mechanical characteristics of hybrid lignocellulosic fiber-reinforced composites are favorable when compared to glass fiber composites, particularly concerning cost and stiffness. In the long term, the utilization of lignocellulosic fibers is emerging as a compelling substitute for synthetic fibers. However, there remains a need for further research to address issues related to natural products, including concerns related to odor and moisture absorption.

References

[1] Campbell, R., "Polyesters," *Encyclopedia of Materials: Science and Technology,* pp. 7171-7172, 2001.

[2] Ebnesajjad, S. and R. A. Morgan, "7 - Fluorinated Additives for Plastics," *Fluoropolymer Additives,* pp. 107-148, 2012.

[3] Hsissou, R., R. Seghiri, Z. Benzekri, M. Hilali, M. Rafik and A. Elharfi, "Polymer composite materials: A comprehensive review," *Composite Structures,* Bd. 262, pp. 2-12, 2021.

[4] Sumrith, N., S. M. Rangappa, R. Dangtungee, S. Siengchin, M. Jawid and C. I. Pruncu, "Biopolymers-Based Nanocomposites: Properties and Application," in *Bio-based Polymers and Nanocomposite*, Springer, Cham, 2019, pp. 255-272.

[5] Saba, N. and M. Jawaid, "A review on thermomechanical properties of polymers and fibers reinforced polymer composites," *Journal of Industrial and Engineering Chemistry,* Nr. 67, pp. 1-11, 2018.

[6] Qin, Q., "Introduction to the composite and its toughening mechanisms," *Toughening Mechanisms in Composite Materials,* pp. 1-32, 2015.

[7] Miskolczi, N., "3 - Polyester resins as a matrix material in advanced fibre-reinforced polymer (FRP) composites," *Advanced Fibre-Reinforced Polymer (FRP) Composites for Structural Applications,* pp. 44-68, 2013.

[8] Jia, H., Y. Sheng, P. Guo, S. Underwood, H. Chen, Y. R. Kim, Y. Li and Q. Ma, "Effect of synthetic fibers on the mechanical performance of asphalt mixture: A review," *Journal of Traffic and Transportation Engineering (English Edition),* Bd. 10, Nr. 3, pp. 331-348, 2023.

[9] Asim, M., M. Jawaid, N. Saba, Ramengmawii, M. Nasir and M. T. H. Sultan, "1 - Processing of hybrid polymer composites—a review," *Hybrid Polymer Composite Materials,* pp. 1-22, 2017.

[10] Nagalakshmaiah, M., S. Afrin, R. P. Malladi, S. Elkoun, M. Robert, M. A. Ansari, A. Svedberg and Z. Karim, "Chapter 9 - Biocomposites: Present trends and challenges for the future," *Green Composites for Automotive Applications,* pp. 197-215, 2019.

[11] Alizadeh-Osgouei, M., Y. Li and C. Wen, "A comprehensive review of biodegradable synthetic polymer-ceramic composites and their manufacture for biomedical applications," *Bioactive Materials,* pp. 22-36, 2019.

[12] Qin, Y., "3 - A brief description of textile fibers," *Medical Textile Materials,* pp. 23-42, 2016.

[13] Efthimiadou, E. K., M. Theodosiou, G. Toniolo and N. Y. Abu-Thabit, "15 - Stimuli-responsive biopolymer nanocarriers for drug delivery applications," *Stimuli Responsive Polymeric Nanocarriers for Drug Delivery Applications,* Bd. 1, pp. 405-432, 2018.

[14] Brink, A. E., *Thermoplastic Polyester Composites,* Newyork City: Hydrosize Technologies, 2004.

[15] Liu, B.-W., Y.-F. Lei, X.-F. Liu, D.-M. Guo, L. Chen and Y.-Z. Wang, "Thermally induced end-group-capturing as an eco-friendly and general method for enhancing the fire safety of semi-aromatic polyesters," *Polymer,* 2021.

[16] Turner, S. and Y. Liu, "5.14 - Chemistry and Technology of Step-Growth Polyesters," *Polymer Science: A Comprehensive Reference,* Bd. 5, pp. 311-331, 2012.

[17] Mochizuki, M., "8 - Synthesis, properties and structure of polylactic acid fibres," *Handbook of Textile Fibre Structure,* Bd. 1, pp. 257-275, 2009.

[18] Rudnik, E., "10 - Compostable Polymer Materials: Definitions, Structures, and Methods of Preparation," *Handbook of Biopolymers and Biodegradable Plastics,* pp. 189-211, 2013.

[19] Amstutz, V., N. Hanik, J. Pott, C. Utsunomia and M. Zinn, "Chapter Four - Tailored biosynthesis of polyhydroxyalkanoates in chemostat cultures," *Methods in Enzymology,* Bd. 627, pp. 99-123, 2019.

[20] Basnett, P., S. Ravi and I. Roy, "8 - Natural bacterial biodegradable medical polymers: Polyhydroxyalkanoates," *Science and Principles of Biodegradable and Bioresorbable Medical Polymers,* pp. 257-277, 2017.

[21] Rudnik, E., "13 - Compostable Polymer Properties and Packaging Applications," *Plastic Films in Food Packaging,* pp. 217-248, 2013.

[22] Jiang, L. and J. Zhang, "7 - Biodegradable and Biobased Polymers," *Applied Plastics Engineering Handbook (Second Edition),* pp. 127-143, 2017.

[23] Luyt, A. S. and S. S. Malik, "16 - Can Biodegradable Plastics Solve Plastic Solid Waste Accumulation?," *Plastics to Energy,* pp. 403-423, 2019.

[24] Visakh, P., 1 - *Polyethylene Terephthalate: Blends, Composites, and Nanocomposites – State of Art, New Challenges, and Opportunities,* William Andrew Publishing, 2015.

[25] Sousa, A. F., C. Vilela, M. Matos, C. S. Freire, A. J. Silvestre and J. F. Coelho, "7 - Polyethylene Terephthalate: Copolyesters, Composites, and Renewable Alternatives," *Poly(Ethylene Terephthalate) Based Blends, Composites and Nanocomposites,* pp. 113-141, 2015.

[26] Turner, S. and Y. Liu, "5.14 - Chemistry and Technology of Step-Growth Polyesters," *Polymer Science: A Comprehensive Reference,* Bd. 5, pp. 311-331, 2012.

[27] Soni, V., V. Pandey, S. Asati, V. Gour and P. K. Tekade, "Chapter 11 - Biodegradable Block Copolymers and Their Applications for Drug Delivery," *Basic Fundamentals of Drug Delivery,* pp. 401-447, 2019.

[28] Zhang, J.-B., X.-L. Wang, Q.-X. He, H.-B. Zhao and Y.-Z. Wang, "A novel phosphorus-containing poly(1,4-cyclohexylenedimethylene terephthalate) copolyester: Synthesis, thermal stability, flammability and pyrolysis behavior," *Polymer Degradation and Stability,* Bd. 108, pp. 12-22, 2014.

[29] Wypych, G., "PTT poly(trimethylene terephthalate)," *Handbook of Polymers (Second Edition),* pp. 595-598, 2016.

[30] Nagarajan, V., A. K. Mohanty and M. Misra, "Reactive compatibilization of poly trimethylene terephthalate (PTT) and polylactic acid (PLA) using terpolymer: Factorial design optimization of mechanical properties," *Materials & Design,* Bd. 110, pp. 581-591, 2016.

[31] Arumugaprabu, V., T. J. Ko, M. Uthayakumar and R. D. J. Johnson, "11 - Failure analysis in hybrid composites prepared using industrial wastes," *Failure Analysis in Biocomposites, Fibre-Reinforced Composites and Hybrid Composites,* pp. 229-244, 2019.

[32] Elfaleh, I., F. Abbassi, M. Habibi, F. Ahmad, M. Guedri, M. Nasri and C. Garnier, "A comprehensive review of natural fibers and their composites: An eco-friendly alternative to conventional materials," *Results in Engineering,* Bd. 19, 2023.

[33] Redwan, A., K. H. Badri and A. Baharum, "An Overview on Lignocellulosic Fibers Ienforced Polymer Composite Materials," *Journal of Al-Nahrain University,* Bd. 20, pp. 25-31, 2015.

[34] Hao, L., S. Sapuan, M. Hassan and R. Sheltami, "2 - Natural fiber reinforced vinyl polymer composites," *Natural Fibre Reinforced Vinyl Ester and Vinyl Polymer Composites,* pp. 27-70, 2018.

[35] Bastidas-Oyanedel, J.-R. and J. E. Schmidt, "Chapter 20 - Waste Biorefinery in Arid/Semiarid Regions," *Waste Biorefinery,* pp. 605-621, 2018.

[36] Shankaran, D. R., "Chapter 14 - Cellulose Nanocrystals for Health Care Applications," *Applications of Nanomaterials,* pp. 415-459, 2018.

[37] Yang, J., Y. C. Ching and C. H. Chuah, "Applications of Lignocellulosic Fibers and Lignin in Bioplastics: A Review," *Polymers (Basel),* 2019.

[38] Chen, H., "Chemical Composition and Structure of Natural Lignocellulose," *Biotechnology of Lignocellulose: Theory and Practice,* pp. 25-71, 2014.

[39] Bajpai, P., "Chapter 2 - Wood and Fiber Fundamentals," *Biermann's Handbook of Pulp and Paper (Third Edition),* Bd. 1, pp. 19-74, 2018.

[40] Rao, X., and J. Barros, "Modeling lignin biosynthesis: a pathway to renewable chemicals," *Trends in Plant Science,* 2023.

[41] Rahman, R., and S. Z. F. S. Putra, "5 - Tensile properties of natural and synthetic fiber-reinforced polymer composites," *Mechanical and Physical Testing of Biocomposites, Fibre-Reinforced Composites and Hybrid Composites,* pp. 81-102, 2019.

[42] Gudayu, A. D., L. Steuernagel, D. Meiners and A. M. Woubou, "Sisal fiber reinforced polyethylene terephthalate composites; Fabrication, characterization and possible application," *Polymers and Polymer Composites,* Bd. 30, 2022.

[43] Kandola, B., S. Mistik, W. Pornwannachai und S. Anand, "Natural fibre-reinforced thermoplastic composites from woven-nonwoven textile preforms: Mechanical and fire performance study," *Composites Part B,* pp. 456-464, 2018.

[44] Sahari, J., S. Sapuan, E. Zainudin and M. Maleque, "Effect of Water Absorption on Mechanical Properties of Sugar Palm Fibre Reinforced Sugar Palm Starch (SPF/SPS) Biocomposites," *Journal of Biobased Materials and Bioenergy,* Bd. 7, pp. 90-94, 2013.

[45] Cherief, M., A. Belaadi, M. Bouakba, M. Bourchak and I. Meddour, "Behaviour of lignocellulosic fibre-reinforced cellular core under low-velocity impact loading: Taguchi method," *The International Journal of Advanced Manufacturing Technology,* Bd. 108, Nr. 12, pp. 223-233, 2020.

[46] Chaitanya, S. and I. Singh, "Sisal Fiber-Reinforced Green Composites: Effect of Ecofriendly Fiber Treatment," *Polymer Composites,* Bd. 39, Nr. 12, pp. 4225-4677, 2018.

[47] Sanadi, A., R. Rowell and R. Young, "Estimation of Fiber-Matrix Interfacial Shear Strengths in Lignocellulosic-Thermoplastic Composites," *MRS Online Proceedings Library,* Bd. 266, pp. 81-92, 1992.

Chapter 6

The Mechanical and Thermal Properties of Polycarbonate Matrix Composites

Laongdaw Techawinyutham*

Department of Production and Robotics Engineering, Faculty of Engineering, King Mongkut's University of Technology North Bangkok, Bangkok, Thailand

Abstract

Polycarbonate (PC) composites comprise a PC matrix reinforced with various materials. Renowned for its transparency, toughness, impact resistance, strength, and stiffness, PC surpasses other transparent thermoplastics like polystyrene (PS) and polymethyl methacrylate (PMMA) in mechanical and thermal properties. These exceptional attributes drive its widespread application. Both natural and synthetic reinforcements, including carbon-based (carbon fiber, nanotubes, graphite, graphene), silica-based (fiberglass, silica particles, aerogel), ceramic (boron nitride, alumina, silicon carbide), and other materials (metals, natural fibers), can be incorporated into PC composites. Extensive research highlights the potential of these reinforcements to significantly enhance the composites' physical, mechanical, and thermal properties, driving their adoption across industries. Carbon-based reinforcements notably improve tensile strength, flexural properties, hardness, and impact resistance while reducing elongation due to increased stiffness. Silica-based and ceramic materials enhance tensile properties and storage modulus but often compromise ductility. Cellulosic-based reinforcements typically decrease tensile strength, elongation, and impact resistance while improving stiffness. Regarding thermal properties, carbon-based and ceramic-filled PC composites

---

* Corresponding Author's Email: laongdaw.t@eng.kmutnb.ac.th

In: Mechanical and Thermal Properties of Engineering …
Editors: Senthilkumar Krishnasamy, Thitinun Ungtrakul, Sabarish Radoor et al.
ISBN: 979-8-89530-460-0

generally exhibit enhanced thermal stability and glass transition temperature (Tg). Silica-based and cellulosic reinforcements can have varying effects on thermal properties.

Keyword: PC composite, synthetic fibers, natural fibers, mineral fillers, carbon-based materials

1. Introduction

Polymer composites are currently garnering significant attention from researchers and manufacturers due to their remarkable properties, including lightweight, high durability, strength, and excellent thermal and electrical insulation. Polycarbonate (PC) is a widely used polymer matrix valued for its high strength, stiffness, impact resistance, and toughness [1].

Several reinforcing materials can be combined with the PC matrix to enhance its physical, mechanical, and thermal properties. Mechanical testing of PC composites typically includes tensile, flexural, impact, hardness, and dynamic mechanical analyses. The influence of reinforcements on crucial thermal properties such as thermal stability, thermal conductivity, glass transition temperature (Tg), and melting temperature (Tm) is also analyzed.

PC composites have been extensively researched and utilized by both academia and industry. To fabricate these composites, a variety of natural and synthetic materials are employed as reinforcement phases within the PC matrix. Carbon-based materials are among the most commonly used reinforcement agents in PC composites (e.g., carbon fiber [2, 3], carbon nanotube [4, 5], graphite [6, 7], and graphene [8, 9, 10, 11]), silica-based materials (e.g., fiber glass [12], silica particle [13], silicon dioxide [13] and aerogel [14]), ceramic materials (e.g., boron nitride [15, 16], alumina [17], and aluminum nitride [18] and silicon carbide [19]), metal and natural materials such as cellulosic fibers and mineral fillers [20, 21, 22, 23, 24, 25].

2. Polycarbonate Composites

PC composites comprise a PC matrix and a reinforcement phase. PC, an amorphous thermoplastic, is synthesized through the reaction of carbonate (-OCOO-) monomers with diols like hexanediol, bisphenol A, or their mixtures [1, 26]. Naturally transparent, PC offers an excellent alternative to glass. It boasts exceptional mechanical properties including extreme toughness, excellent impact resistance, high strength, and stiffness. Compared to other transparent thermoplastics like polystyrene (PS) and polymethyl methacrylate (PMMA), PC exhibits superior impact resistance, elongation, and creep resistance [1, 27]. PC possesses a high Tg of 145°C, a Tm range of 225-250°C, and an HDT of 135°C under 0.46 MPa. These thermal properties surpass those of PS and PMMA, which have Tg of 100°C and 105°C, respectively. Additionally, PC's HDT and thermal conductivity exceed those of PMMA [1].

PC is used in numerous applications due to its excellent mechanical and thermal properties. These include packaging (e.g., medical device film), building and construction (e.g., laminated glass, bulletproof glass, lamp housings), automotive components (e.g., light signal lenses, head/rear light housings, headlamp lenses, rooftop panels), electronic components (e.g., switch covers, battery covers, relay covers), appliances (e.g., coffee maker water catchment, food mixer housing), and safety equipment (e.g., safety goggles, machine guards, safety shields, helmets) [1, 28].

The physical, mechanical, and thermal properties of PC can be enhanced by incorporating additives, fillers, or fibers to create polymer composites with tailored properties. Polymer composites comprise at least two distinct phases: a polymer matrix and a reinforcement phase consisting of fillers and/or fibers [29, 30]. Reinforcement materials in polymer composites can include metals, ceramics, and other polymers [29]. By combining materials with differing properties, composites exhibit exceptional characteristics like lightweight, high strength-to-weight ratio, and high stiffness, making them suitable for advanced applications in construction, sports, medical equipment, aerospace, and automotive industries [29, 31].

PC composites comprise PC as a matrix and reinforcement materials that can be natural or synthetic. The most common reinforcement agents incorporated into the PC matrix are synthetic materials such as nanosilica [13], carbon nanotube [5], carbon fiber [3], fiber glass [12], and graphene [9], as well as some natural materials of cellulosic fibers [22, 23], etc. [1, 28].

3. Reinforcement Materials

The types of reinforcement materials used in polymer composites and the quality of the bond between the reinforcement and polymer matrix are critical in determining the physical, mechanical, and thermal properties of PC composites. A variety of materials, including carbon-based, silica-based, metallic, ceramic, cellulosic, and mineral fillers, can be employed as reinforcements within the PC matrix.

3.1. Carbon-Based Materials

The most commonly used synthetic reinforcement materials in PC matrices are carbon-based materials, including carbon powder, carbon fiber, carbon nanotubes, graphite powder, graphite flakes, and graphene [32, 33]. These materials are prized for their exceptional mechanical and thermal properties, high thermal conductivity, and low density. Incorporating carbon-based materials into PC matrices significantly enhances the mechanical properties of the resulting composites. Carbon-based material/PC composites have been extensively studied for critical applications in aerospace, automotive, military, defense, turbine blades, construction, pressure vessels, offshore tethers, medical, and sports equipment [32, 34, 35, 36, 37, 38].

Carbon fiber, composed of at least 92 wt% carbon, and graphite fiber, containing 99 wt% carbon, are prominent carbon-based materials. Carbon fiber is renowned for its exceptional tensile properties, physical attributes, low density, excellent thermal and chemical stability, high thermal and electrical conductivity, and superior creep resistance, making it a highly sought-after reinforcement in polymer matrices [37, 38]. Carbon fiber is available in various forms, including woven textiles, prepregs, continuous fibers/rovings, and chopped fibers [37].

Carbon nanotubes (CNTs), with an outer diameter ranging from about 2.5 to 30 nm, have potential mechanical, thermal, and electronic properties. The structure of carbon nanotube is the rolled-up layer of carbon atoms bonded together in a hexagonal mesh as shown in Figure 1 [32, 39]. There are two types of carbon nanotubes: single-walled carbon nanotubes (SWCNTs) and multi-walled carbon nanotubes (MWCNTs). SWCNTs consist of a single layer of carbon atoms, while MWCNTs consist of several layers of carbon atoms [32]. The diameter of SWCNTs is smaller than that of MWCNTs [39]. Incorporating carbon nanotubes into the polymer matrix is of great interest due to their remarkable properties [40].

Graphite is a natural mineral with an anisotropic crystalline structure, consisting of hexagonal layers of carbon atoms arranged in a specific pattern, as shown in Figure 1 [41, 42]. It possesses excellent thermal and electrical conductivity and can be used as a reinforcement material in a polymer matrix to improve its flexibility, thermal and electrical conductivity, although it reduces the ductility of the resulting nanocomposites [42, 43].

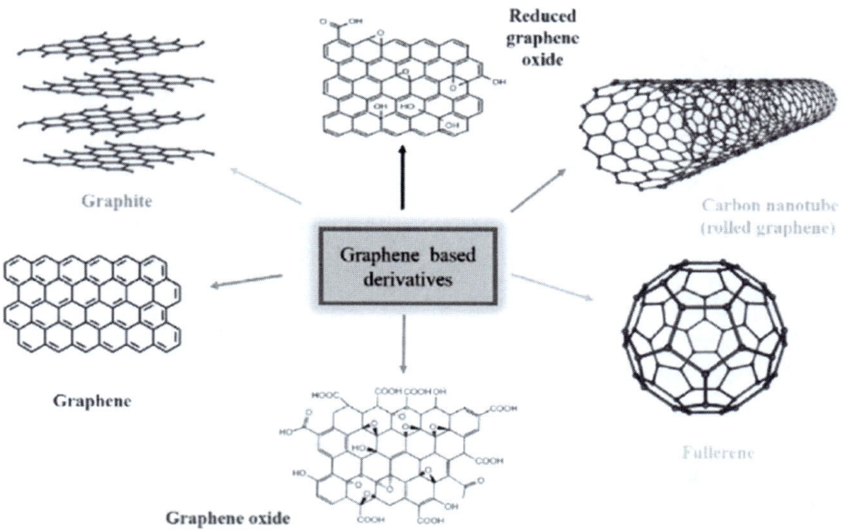

Figure 1. Systematic diagram of carbon-based materials. Reprinted from work of A. Hashmi et al. [45].

Graphene, a monolayer carbon nanostructure with a hexagonal pattern similar to a honeycomb structure, is known for its remarkable mechanical and thermal characteristics, lightweight, transparent, and good heat and electricity conduction, as illustrated in Figure 1 [10, 40]. The outstanding properties of graphene, such as high specific surface area, high aspect ratio, and notable thermal and mechanical properties, have generated a lot of interest in graphene-based polymer composites for improving the properties of nanocomposites [9]. The popular method of producing graphene or reduced graphene oxide is by reducing graphite oxide or graphene oxide (GO) to reduced graphene oxide (RGO) [44]. The production of graphene is based on the exfoliation of graphite [44]. There are several types of graphene based on the production method and characteristics, such as graphene, graphene nanosheets, graphene oxide, exfoliated graphene oxide, microwave-exfoliated graphene oxide, reduced graphene oxide, thermally reduced graphene oxide, and chemically reduced graphene oxide [44]. It is important to note that graphite nanoplate is frequently incorrectly called by producers as graphene or graphene nanoplatelets [44]. Using graphene as a reinforcement material in a polymer matrix can improve the electrical, mechanical, and thermal properties of PC composites [44].

3.1.1. Silica-Based Materials

The silica-based materials, which are major components of glasses and ceramics, are commonly used as reinforcement materials in polymer matrices. The silica-based materials such as fiber glass, silicon dioxide, nanosilica particles, and sol-gel solutions are frequently used due to their excellent thermal stability and mechanical properties [46, 47].

Fiberglass contains more than 50% silica content in the form of thin glass fibers with a diameter ranging from 3–15 μm [48, 49]. There are several types of fiberglass, such as short fiber, long fiber, woven and nonwoven fabrics, which are extremely interesting reinforcements in polymer matrix due to their high resistance to chemicals and heat, and good mechanical properties [47, 48, 49].

Silica aerogel consists of nanoporous silica particle networks and high-volume nanopores arranged in a pearl necklace-like structure. It is characterized by its low density, high specific surface area, good transparency, and excellent thermal and electrical insulation properties [14, 46].

3.1.2. Ceramic Materials

Ceramic-based materials with good thermal conductivity are often used as reinforcement in polymer matrix composites. This is because polymers typically have low thermal conductivity, but adding ceramic fillers can create thermally dissipative paths in the matrix, resulting in improved density, thermal conductivity, electrical insulation, and mechanical performance [15, 16, 50]. These composites find the use in high-temperature applications, electric stress control, electromagnetic shielding, and energy storage, such as in battery units for electronics and alternative fuel vehicles [16, 34, 51]. Common ceramic fillers reinforced in PC include boron nitride, alumina (Al_2O_3), aluminum nitride, silicon nitride, and silicon carbide [15, 34, 51].

3.1.3. Metallic Materials

The high strength/modulus values, good thermal stability, thermal conductivity, and recyclability of metals contribute to the high usage of metallic micro- and/or nanoparticles as inorganic fillers to reinforce polymeric materials [52, 53]. Commonly employed metallic powders, such as iron, tungsten, silver, copper, aluminum, and zinc, can enhance polymer composite properties, including mechanical properties, electrical conductivity, thermal conductivity, and gas barrier properties, and are applied in potential applications [52, 54].

3.1.4. Natural Materials

Natural materials, such as cellulosic fibers from plants, animal-based materials, and mineral fillers, are extensively used as reinforcement materials for polymer composites due to their low cost and low environmental impact [21, 55, 56, 57, 58]. The natural fibers from plants such as bamboo, wood, sisal, water hyacinth, coconut, hemp, pineapple leaf, and cotton fibers, which are derived from agricultural and processing wastes, have been extensively used as reinforcement materials in various polymers [55, 59, 60]. Cellulosic fibers are currently interesting for fabricating polymeric materials due to their biodegradability, lightweight, renewability, reasonable cost, and eco-friendly [55, 61]. However, cellulosic fibers have limitations such as high moisture absorption and poor interaction with the polymer matrix [61].

Natural mineral fillers are inorganic materials derived from natural minerals such as titanium dioxide (TiO_2), calcium carbonate ($CaCO_3$), talc, basalt, dolomite, and mica [21, 56, 57, 58, 62, 63]. Mineral fillers are excellent reinforcement materials for polymer composites due to their low cost, high rigidity, strength, stiffness, hardness, as well as good abrasion, thermal, and chemical resistance. It can effectively increase mechanical, physical, and thermal properties and reduce the cost of polymer composites [56, 58]. Moreover, mineral fillers such as titanium dioxide can be used as pigments for polymeric materials [57].

4. Mechanical Properties of PC Composites

The mechanical properties, such as tensile properties (e.g., tensile strength, tensile modulus, and elongation), flexural properties (e.g., flexural strength and flexural modulus), impact resistance, hardness, and dynamic mechanical properties, are important material properties that need to be tested for potential applications. Various studies commonly describe the mechanical behavior of materials as the reaction of any material during loading leading to deformation [64]. The results of mechanical testing are often used to design parts for specific applications, to control the quality of materials, and to investigate the influence of reinforcement material, size, and loading. The mechanical properties of polymer composite materials are significantly affected by the mechanical properties of the polymer matrix, mechanical properties of the reinforcement phase, concentration of the reinforcement, size of the reinforcement, dispersion of the reinforcement in the matrix, and matrix-reinforcement interaction, among others [65, 66].

The mechanical properties of PC composites can be significantly improved by incorporating various types of reinforcement. Reinforcement phases can take several forms, such as particles, fibers, and fiber fabrics, which can be made from natural or synthetic materials. Common reinforcement materials for PC matrix composites include carbon-based materials, silica-based materials, metals, ceramics, cellulosic fibers, and mineral fillers. Numerous studies have investigated the effects of these reinforcement materials on the mechanical properties of PC composites, considering factors such as the type, concentration, size, and dispersion of the reinforcement phase, as well as the matrix-reinforcement interaction.

4.1. Carbon-Based Reinforcement

The carbon-based materials such as carbon fiber, carbon nanotube, graphite, and graphene are commonly used as a reinforcement into the PC matrix because of their remarkable mechanical performances [32, 33].

There are many types of carbon fiber used to reinforce PC matrix, such as carbon fiber fabric, long and short carbon fiber. Numerous studies have investigated the mechanical properties of carbon fiber/PC composites. One study involved arranging 6 plies of continuous carbon fiber (non-crimped fabrics) in the same unidirectional (UD) laminate with PC [3]. The laminate had a thickness of 1.14 mm and consisted of carbon fiber at 48vol%. The mechanical properties were tested in two directions, 0° and 90°, and both directions showed better mechanical results than those of neat PC. The tensile strength and tensile modulus of the laminate composites in the fiber direction (0°) exhibited the best tensile properties, with improvements of around 1075% and 2230% over pure PC, respectively. Furthermore, the flexural properties in the fiber direction showed improvements in flexural strength and flexural modulus of the carbon fiber/PC composite by 415% and 2130%, respectively, compared to neat PC. In another study, the incorporation of four layers of carbon fiber fabric (20.4 wt%) in PC developed a tensile modulus corresponding to a 57.5% enhancement compared to neat PC [67]. The addition of carbon fiber to PC matrix resulted in a significant increase in the hardness, tensile modulus, and storage modulus of the composites at carbon fiber concentrations of 10, 15, and 25wt%. This improvement was due to the well dispersion of the carbon fiber in the polymer matrix and strong interfacial bonding between the carbon fiber and PC matrix [68]. However, when carbon fibers were added to a PLA/PC blend, the tensile strength and storage modulus of the composites were enhanced, but the elongation at break decreased. This reduction in elongation was due to the addition of high stiffness carbon fiber and the occurrence of microcracks at the end of the fibers in the matrix. The highest tensile strength value and storage modulus were achieved with the addition of 30wt% carbon fiber [69]. C. Ozkan et al. [2] have also confirmed that the tensile strength and tensile modulus of PC were improved by 104% and 278%, respectively, with the addition of 6 cm length carbon fiber, as compared to neat PC. However, the impact resistance of the PC composites was not significantly affected by the addition of carbon fiber. Moreover, the mechanical properties of PC composites can be improved by adding carbon nanofiber. The storage modulus, loss modulus, and flexural properties of PC composites significantly increased with small loading of carbon nanofiber (0-1 wt% [70] and 0.5-7 wt% [71]). The highest enhancement in the storage modulus was achieved with an improvement of 21% [70] and 44% [71] as compared to pure PC, due to the very high surface area of the carbon nanofiber and good filler-polymer matrix interfacial adhesion, which allowed for the transmission of applied stress from the PC matrix to the carbon nanofiber. Carbon nanofiber can also help retain modulus at higher temperatures [71].

The use of PC as a matrix phase for CNT can improve the mechanical properties of CNT/PC composites [39]. Even a small amount of CNT at 0.3 wt% incorporated in carbon fiber/PC composites can improve tensile strength, tensile modulus, impact strength, and storage modulus (evaluated by DMA) with an increment of 21.8%, 7.5%, 33.3%, and 18.5% compared with carbon fiber/PC composites, respectively. This is due to the outstanding properties of CNT and the electrophoretic deposition treatment covering CNTs on the carbon fiber surface, which enhances the carbon fiber-PC matrix interaction and decreases stress concentration between carbon fiber and PC matrix [72]. The tensile strength of CNT/PC composite was improved from 75.2 MPa (neat PC) to 79.6 MPa with the addition of 0.5 wt% CNT; however, elongation reduced from 9.6% to 7.8% [73]. The surface-modified carbon fiber-CNT reinforced in PC can dramatically increase the mechanical properties of the PC composite.

The MWCNTs were added to polymeric blends and/or other reinforcement materials to improve the mechanical properties of MWCNT/PC composites. The tensile modulus, tensile strength, and storage modulus of the composites increased with an increasing weight percentage of MWCNTs, while elongation decreased [4, 74, 75, 76, 77]. Addition of 2 wt% MWCNTs efficiently improved the tensile modulus by up to 89%. The improvement in mechanical properties was observed even at low concentrations of MWCNTs in PC, and the mechanical properties further increased with an increase in the MWCNT composition due to the good stress transfer provided by the large surface area of MWCNTs and good filler-matrix interfacial adhesion [4]. However, the insertion of MWCNT resulted in a reduction in the damping behavior and storage modulus due to the expansion of free volume of the polymer in the rubbery state at high temperature testing, which mobilized the polymer chain movement [77]. In another study, MWCNTs at concentrations of 0.5, 1, 3, 5, and 10 phr were composited with an ethylene methyl acrylate (EMA)/PC blend at a content of 5/95 wt% and [78]. The storage modulus of the composites evaluated by DMA enhanced with increasing MWCNT loading, resulting from the improvement in interfacial adhesion between MWCNT and the polymer matrix. The incorporation of MWCNTs at 10 phr in EMA/PC blend significantly developed storage modulus and loss modulus by 35.1% and 88.8%, respectively, compared to neat EMA/PC blend since the increment of friction can improve the energy transfer. On the other hand, the damping factor (tan δ) of the composites decreased with an increase in MWCNT content and sharply dropped by 36% at 10 phr of MWCNT compared to the neat blend, since MWCNT filler obstructed the molecular chain mobility of polymer, resulting in an enhancement of stiffness [78]. The tensile strength, tensile modulus, and elongation at break were found to decrease with increasing MWCNT content (0.1, 0.5, and 1 wt%) [79]. However, the addition of only 1 wt% MWCNT in carbon fiber fabric/PC composite dramatically improved the storage modulus and impact resistance since MWCNT hindered the polymer chain motion [80, 81]. In addition, the tensile strength and storage modulus of MWCNT/polyamide 66 (PA66)/PC composites significantly increased with increasing MWCNT content compared to PA66/PC blend due to the strong compatibility effect of the hybrid fillers in the polymeric blends. When 10 wt% MWCNT was added to EMA/PC blend, the tensile strength and tensile modulus of the composites were superior to neat EMA/PC blend, but the composites had lower impact resistance and elongation at break than the EMA/PC blend. However, the impact strength of all EMA/PC composites was better than neat PC [82]. Furthermore, the tensile strength and tensile modulus of the composites enhanced but elongation at break reduced with an increased loading of MWCNTs in EMA/PC blend [74]. The addition of MWCNT at only 5 phr and 10 phr can improve tensile strength up to 37% and 39%, respectively, compared to EMA/PC blend. Moreover, the maximum tensile modulus presented at 10 phr content of MWCNT was 60% higher than unfilled EMA/PC blend due to good load transferring from PC matrix to the MWCNTs [74]. However, the impact strength of the composites increased with the increase in the loading of MWCNTs (0.25-3 phr), but higher MWCNT concentrations (4-10 phr) decreased impact strength due to the accumulation of MWCNT (poor dispersion) in the PC matrix [74].

The tensile strength and tensile modulus of 10 wt% MWCNT/PC/EMA composites were found to be higher than those of 10 wt% graphene/PC/EMA composites by 8.3% and 11.1%, respectively [82]. Moreover, the incorporation of hybrid fillers, i.e., both MWCNT and graphene, in the matrix contributed to higher tensile strength, tensile modulus, and impact resistance compared to the graphene/PC/EMA composite and MWCNT/PC/EMA composite. This can be attributed to the increase in stiffness and polymer chain restriction due to the addition of MWCNT and graphene, well dispersion of the nanofillers in the matrix, the linking between MWCNT and graphene layers, and strong interfacial interaction between the nanofillers and polymer matrix [82].

Furthermore, the addition of graphite to the PC matrix resulted in a significant improvement in the stiffness of the composite, with an increase in tensile modulus and flexural modulus of around 150% and 60%, respectively, compared to neat PC. This is due to the homogenous dispersion of graphite in the PC matrix [7].

The addition of graphite flakes (10%, 20%, and 30%) in PC matrix improved the tensile strength, tensile modulus, and storage modulus of the composites due to strong interfacial adhesion between the graphite flakes and the PC matrix, allowing for effective stress transfer [6]. In contrast, the tensile strength and tensile modulus of MWCNT/PC composites decreased with an increase in MWCNT content and were lower than the graphite flake/PC composites. This was due to the agglomeration of MWCNT within the PC matrix, which hindered their ability to effectively reinforce the matrix [6]. The presence of graphite flakes resulted in decreased impact strength and elongation of the PC matrix, as the mobility of polymer chains was restricted by the presence of the graphite flakes, leading to increased brittleness. Although the mechanical properties of graphite flake/PC composites were superior to those of MWCNT/PC composites, the presence of MWCNT at a content of 20 wt% in PC led to better storage modulus than the insertion of 30 wt% graphite flake. This is because the inherently high modulus of MWCNT can create a strong intermolecular bonding between the filler and PC matrix [6].

The tensile strength of PET/PC blend reinforced with graphite nanoplatelets was higher than that of PET/graphite nanoplatelets composites and PC/graphite nanoplatelets composites due to the transesterification reaction acting as a compatibilizer that improved the interfacial adhesion between PBT and PC, resulting in better load transfer. PC has higher tensile strength than PBT [83]. The addition of different graphite nanoplatelets at 10 wt% to PC matrix increased tensile modulus but decreased the tensile strength and elongation at break compared to pure PC due to poor distribution of graphite nanoplatelets in the PC matrix. The different structures of graphite nanoplatelets, such as lamellar structure, fragmented structure, and compact structure, significantly affects the tensile properties of the PC composites. The best tensile properties of PC composites were observed with the addition of graphite nanoplatelets with a lamellar structure, and has an improvement of 160% over pure PC, which was also better than CNT/PC composites with an improvement of 130% over pure PC. The tensile properties of PC composites with graphite nanoplatelets in fragmented structure were higher than those of PC composites with graphite nanoplatelets in compact structure, as seen in Figure 2 [43].

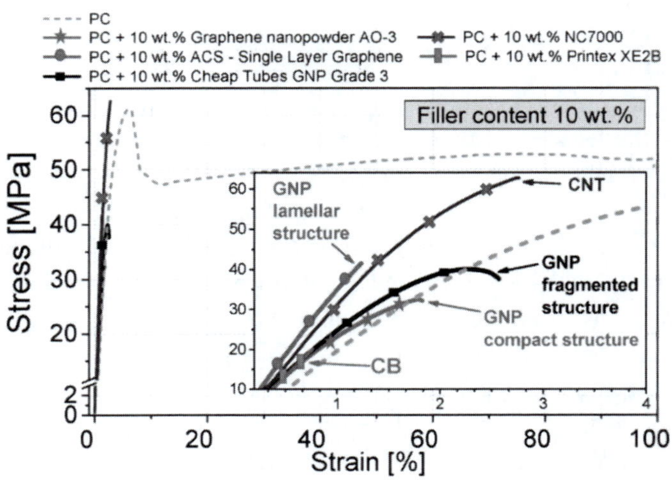

Figure 2. Stress–strain graph of carbon-based materials/ PC composites. Reprinted from work of M. T. Müller et al. [43].

PC composites reinforced with exfoliated graphite nanoplatelets at loading levels of 2-15 wt% exhibited higher tensile and flexural moduli compared to carbon fiber/PC and carbon nanotube/PC composites. These properties were found to increase with increasing exfoliated graphite nanoplatelets loading [84]. The dynamic mechanical properties of microwave-exfoliated graphite oxide/PC composites at a loading level of 3 wt% showed a 30% increase in storage modulus over neat PC at near room temperature, and a 400% increase in storage modulus in the rubbery state. However, the magnitude of tan delta and impact strength of the composites decreased with increasing graphite oxide content [85].

The inclusion of graphene nanoplatelets in a polymer matrix gradually improves the mechanical properties, especially the stiffness of the polymer composites. Additionally, the mechanical strength increases as a function of graphene nanoplatelet concentration, showing a similar trend with the incorporation of CNTs. The addition of only 0.5 wt% of graphene nanoplatelets enhances the tensile yield strength by 24% compared to neat PC [10]. The addition of graphene nanoplatelets to the PC composite acts as a reinforcement but has no effect on the modulus. However, elongation at break decreases with the addition of 0.5 wt% [10]. Furthermore, a higher addition of graphene nanoplatelets at 10 wt% shows a 33% higher tensile modulus than neat PC due to the mobility restriction of polymer chains resulting from the large aspect ratio of graphene nanoplatelets and their interaction with the PC matrix. However, strain reduces with an increase in graphene nanoplatelet concentration [9].

The addition of graphene nanoplatelets (1, 3, and 5 wt%) in the polymer blend has a significant reinforcing effect on the mechanical properties of acrylonitrile butadiene (ABS)/PC blend (30/70 composition) [11]. The maximum tensile strength, elongation at break and flexural strength were observed with the addition of 3 wt% of graphene nanoplates in the ABS/PC blend, which showed a 15%, 94%, and 11% enhancement as compared to the neat ABS/PC blend, respectively. The improvement in stiffness of the nanocomposites without the loss of ductility is due to the homogenous dispersion of graphene nanoplates in the PC/ABS matrix and the strong interaction between the nanofillers and PC/ABS [11]. However, the maximum tensile strength, elongation at break and flexural strength decreased with the incorporation of 5 wt% graphene nanoplates in the ABS/PC blend due to the agglomeration of graphene nanoplates [11]. In another study, the insertion of 5 wt% graphene nanoplates into PC/ABS blend showed a 91% increase in tensile modulus over unfilled PC/ABS blend due to the homogenous dispersion of graphene nanoplates in the PC/ABS matrix and the good interaction between graphene nanoplate and polymer matrix [8]. The presence of 1 wt% graphene nanoplates in the PC/ABS blend resulted in a dramatic decrease in impact strength (around 91% relative to the unfilled blend) since the energy absorbing mechanism of the nanocomposites changed and acted as crack initiators in the matrix [11]. However, the hybrid reinforcement of graphene nanoplatelet at 1 wt% and carbon nanotube at 1 wt% significantly enhanced the tensile yield strength by 42% over neat PC [16]. The combination of functionalized graphene sheets and graphite particles reinforced in the PC matrix did not show significant improvement in tensile modulus and dimensional stability [7]. Moreover, graphene can increase the wear resistance of the composites since the graphene-PC interlocking was enhanced by the wrinkled and rough surface of graphene, providing the load transfer [86].

The presence of graphene flakes has been shown to increase both tensile yield strength and tensile modulus in PC composites [87]. Specifically, the addition of 1 wt% graphene flakes to the PC matrix results in a 26% increase in tensile modulus, although further increases in graphene loading result in a reduction of tensile modulus [87]. Even at a loading of 1.5 wt%, the addition of graphene flakes still results in a 17% improvement in tensile modulus compared to neat PC [87].

4.2. Silica-Based Materials

Silica-based materials/PC composites have been widely studied in research groups due to their outstanding mechanical performance. These composites exhibit remarkable mechanical behavior in terms of tensile strength, compressive strength, stiffness, and electrical properties, but relatively low impact resistance. The addition of glass fiber to the polymer matrix contributes to high stiffness, good mechanical and tribological properties but low ductility [48]. The incorporation of glass fiber significantly alters the mechanical properties of the composites [88]. The tensile strength and modulus of the glass fiber/PC composites are higher than those of PC due to the increase in the content of high modulus glass fiber in the PC matrix [88]. The tensile strength was improved by 29.15%, 58.25%, and 73.38% as the amount of glass fiber in the specimens increased to 10%, 20%, and 30%, respectively. The composition of glass fiber at 40wt% in PC after aging at 100°C for 8 hours provided the maximum tensile and flexural strength, as the aging improved the bonding at the fiber/polymer interface [12]. Incorporating 10 wt% of glass fiber into a PBT/PC blend resulted in an effective enhancement of tensile strength by 61.45%, flexural strength by 82.05%, and impact by 64.35% over the virgin PBT/PC blend due to the higher strength of glass fiber compared to that of the polymer matrix [89]. The hardness values of the glass fiber/PC composites improved as a function of fiber content, with an enhancement by 24.3%, 28.6%, and 30% with the insertion of glass fiber at 10%, 20%, and 30%, respectively [88]. However, the ductility of the composites decreased with the incorporation of fiberglass, which presented lower impact strength and elongation at break compared to the neat polymer matrix. The low ductility of glass fiber improved the stress concentration at the ends of the fiber glass in the matrix [88, 89].

The nanosilica fillers were reinforced in PC matrix. The increase in nanosilica concentration enhanced the tensile properties and storage modulus; however, the addition of nanosilica higher than 0.6 wt% reduced the tensile modulus, yield stress and storage modulus. This mechanical improvement occurred due to the essential stiffness of nanosilica, organized dispersion of nanosilica particles in the polymer matrix, and good reinforcement-matrix interfacial adhesion [67]. Moreover, the addition of 0.6 wt% nanosilica to the laminate layers of the carbon fiber fabric/PC composite dramatically enhanced the tensile modulus by 94% compared to neat PC. Additionally, compared to the carbon fiber/PC composite, the tensile modulus increased by 92%, the yield stress by 55%, and the storage modulus by 19%. However, the elongation of the PC composite decreased. [67].

The addition of silicon dioxide (SiO_2) nanoparticles (with diameters of 7 and 12 nm) to PC matrix led to a significant improvement in impact strength, particularly for the smaller 7 nm particle size. The smaller SiO_2 particles improved impact strength and fracture toughness by 246% and 27%, respectively, compared to PC. This improvement is due to their larger specific surface area, which provides strong bonding between the nano-SiO_2 and the PC matrix, dissipating filler-matrix energy transfer. Additionally, the smaller fillers enhanced transparency [13].

In a PMMA/PC blend (in a ratio of 25/75), the tensile strength and modulus enhanced with increasing silica aerogel concentration, but the impact strength and elongation at break decreased. This is because the crosslinking silica networks in silica aerogel, with a high aspect ratio, increased the rigidity of the silica aerogel compared to neat PC and PMMA, restricting the polymer chain movement. The agglomeration of silica aerogel reduced filler-matrix load transfer, leading to a reduction in impact strength and elongation at break.

The incorporation of silica aerogel at only 5 phr provided the maximum tensile properties [14].

4.3. Ceramic-Based Materials

Ceramic-based fillers with high mechanical performance includes boron nitride, alumina (Al_2O_3), aluminum nitride, silicon nitride, and silicon carbide [15, 34, 51]. Boron nitride is a widely used ceramic material combined with PC matrix, which increases the stiffness but decreases the toughness of PC composites. The tensile yield strength, flexural strength, elongation at break, tensile fracture toughness, and impact strength of boron nitride/PC composites decrease with increasing boron nitride concentration [15, 16]. The incorporation of 5 wt% boron nitride in the PC matrix decreased the tensile yield strength and elongation at break of the boron nitride/PC composite by 8% and 34%, respectively, compared to neat PC [16]. However, the mechanical properties of the boron nitride/PC composites in terms of elongation at break, tensile fracture toughness, and impact strength were significantly improved by the addition of 1 wt% graphene nanoplatelets and 1 wt% carbon nanotubes [15, 16]. The tensile strength of boron nitride/PC composites decreased with increasing boron nitride content due to network formation (aggregation) of boron nitride within the PC matrix and weak filler-matrix interfacial interactions that lead to the creation of micro voids. The tensile modulus of boron nitride/PC composites increased with increasing filler content due to better distribution of boron nitride in the PC matrix and the reinforcement effect of the added boron nitride fillers [34]. The addition of a small amount of boron nitride to the PC matrix can improve the tensile strength of PC composites since boron nitride terminates crack growth; however, a high loading of boron nitride reduces tensile strength since the excessive concentration of boron nitride causes stress concentration in the PC matrix [90]. The increase in boron nitride content decreases the elongation at break and impact strength of the PC composite [50, 90]. The storage modulus significantly enhances with the increasing boron nitride loading, which is much higher than that of neat PC [50]. This is because the composite needs to store enough energy before the molecular chain can move [50]. Boron nitride nanotubes at 1 wt% loading in the PC matrix can improve tensile toughness by 39% over neat PC; moreover, the tensile stress and tensile modulus have an improvement of 12% and 13% (1 wt% of boron nitride nanotubes) and 14% and 31% (4 wt% of boron nitride nanotubes), respectively [91]. The transparency of boron nitride nanotubes/PC composite thin films decreases with increasing filler loading. The storage modulus increases with an increased boron nitride nanotube content, which has an improvement of about 8% at 60 °C, while elongation and impact strength decrease with an increase in boron nitride nanotube loading [91]. Boron carbide at 5 vol% into the PC matrix can improve the hardness and wear resistance of PC composites due to the high hardness of the filler [86].

The tensile strength of 3D specimens of alumina/PC nanocomposites was found to be 109% higher than that of pure PC; however, the flexural strength decreased due to the crack growth from shrinkage holes and voids in the matrix caused by the presence of alumina [17]. The impact strength and hardness of the alumina/PC nanocomposites were found to be higher than those of neat PC by 111% and 11%, respectively, due to crack deviation of hard alumina particle [17]. The use of large alumina particles as reinforcement in PC showed better mechanical properties than those of small alumina particles, which is attributed to the small filler network structures formed by the higher specific surface area of the smaller particles [15]. Increasing the alumina loading in the PC matrix did not affect the tensile strength and flexural strength, but it reduced the impact resistance and elongation of the composite [15].

The research on PC composites with ceramic materials such as aluminum nitride and titanium nitride is limited. The addition of 2 wt% of aluminum nitride to PC matrix provided the best mechanical properties, including tensile strength and tensile toughness, which were improved by 32.8% and 51.6%, respectively, compared to neat PC [18]. However, the addition of higher concentrations of aluminum nitride reduced the mechanical properties, and the storage modulus of the aluminum nitride/PC composites than that of pure PC [18].

On the other hand, the introduction of titanium nitride nanoparticles into the PC matrix increased tensile properties, flexural properties, and storage modulus. The maximum improvement in mechanical performance was observed at 3 wt% of TiN, with enhancements of 35.5%, 29%, 13.6%, and 3.7% for tensile strength, tensile modulus, flexural strength, and flexural modulus, respectively, compared to pure PC [31]. However, the impact strength increased at low filler loadings but then decreased as a function of nanoparticle loading. The highest improvement in impact strength was observed at 2 wt% of TiN, with an improvement of 16.2%, and the maximum hardness was exhibited at 6 wt% of TiN, with an enhancement of 20.2% compared to neat PC. Low concentrations of titanium nitride in the PC composites decreased the microhardness value [31].

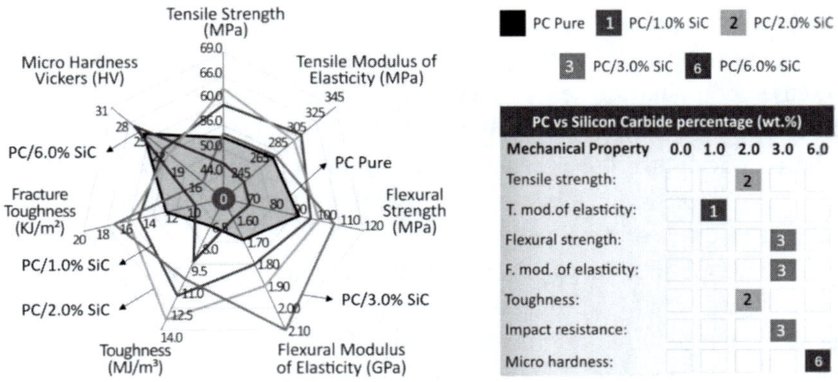

Figure 3. Overview of the mechanical results of silicon carbide/PC composites. Reprinted from work of M. Petousis et al. [19].

The addition of silicon carbide nanoparticles to PC can improve the mechanical performance. Specifically, at a concentration of 2 wt%, the tensile strength of the composite was found to increase by 19.5%. However, the maximum loading of silicon carbide at 6 wt% reduced the tensile strength due to reaching the percolation threshold [19]. The addition of only 1 wt% of silicon carbide resulted in an improvement of the tensile modulus by 9.6% over PC, likely due to an increase in stiffness [19]. The highest impact resistance, flexural strength, and flexural modulus were observed at the addition of 3 wt% of silicon carbide, with improvements of 35.9%, 16%, and 25.4%, respectively. These properties then decreased with increasing filler content at 6 wt% due to the saturation content of silicon carbide, as shown in Figure 3 [19]. However, the storage modulus of silicon carbide/PC composites was lower than that of neat PC, likely due to the elastic behavior of silicon carbide nanoparticles [19].

4.4. Metallic Materials

Metal materials possess high mechanical properties and can be used as reinforcement in the polymer matrix to improve its mechanical performance. However, metal is not commonly used as a reinforcement phase, leading to limited research on PC composites with metal fillers. When tungsten microparticles were added at 1%, 3%, and 5% by weight to PC for printing 3D specimens, a constant tensile strength value was observed, while elongation at break decreased, and impact resistance and storage modulus were enhanced as observed by DMA [92].

4.5. Other Natural Materials

Natural materials and waste products such as cellulose, wood sawdust, pineapple leaf fiber, fly ash, silk waste, and cotton waste have been used as reinforcement phases in PC matrices, but the research on these materials is limited. Cellulose nanocrystals at a concentration of 3 wt% were dispersed well in PC composite films, resulting in good transparency, high tensile strength, and high tensile modulus with an improvement of 30.6% and 27.3%, respectively. However, the elongation at break of the composite films decreased due to the restriction of polymer chain mobility [22]. The transparency, tensile strength, and tensile modulus decreased with further increases in cellulose nanocrystal content due to cellulose nanocrystal agglomeration [22].

The tensile modulus of wood sawdust/PC composites (at 10, 20, and 30 wt% of wood sawdust) was higher than that of pure PC by 20.5%. The rigid filler particles of wood sawdust enhanced stiffness and immobilized the polymer chain [23]. The tensile modulus of wood sawdust/PC composites improved, but the tensile strength dropped as the wood sawdust content increased due to the agglomeration of wood particles and poor filler-matrix interfacial interaction, which resulted in small gaps between filler and matrix decreasing load transfer with increasing filler loading [23].

The tensile modulus increased as the pineapple leaf fiber content increased in the PC matrix (0, 5, 10, and 20 wt%). The highest tensile modulus was observed when the surface of the pineapple leaf fiber was modified by sodium hydroxide. However, the maximum tensile strength and impact strength were obtained when the fibers were treated with silane, as this treatment improved the interfacial interaction between the pineapple leaf fiber and the PC matrix, providing good stress transfer from the matrix to the fiber [93]. The incorporation of untreated pineapple leaf fibers resulted in a steady tensile modulus, but tensile strength slightly decreased. Moreover, elongation and impact strength decreased with increasing pineapple leaf fiber content, as the fibers hindered the chain mobility of the polymer matrix [93].

Silk and cotton fibers are commonly used in clothing and textiles, but their use as a reinforcement phase in polymer composites has been limited in research. The good physical and mechanical properties of silk fibers make them a promising candidate to enhance the mechanical and physical properties of polymer composites [94]. M Taşdemır et al. [94] found that the incorporation of silk waste to the PC matrix resulted in better tensile modulus, elongation, and hardness compared to the addition of cotton waste to PC, but the tensile strength, elongation, and impact resistance of cotton waste/PC composites were higher than those of silk waste/PC composites. The tensile strength, tensile modulus, and hardness decreased, which were lower than that of neat PC, but elongation and impact resistance increased with increasing cotton loading in PC, which were higher than neat PC. The tensile strength, elongation, and impact of silk waste/PC composites were lower than neat PC, but tensile modulus and hardness increased with increasing silk loading, which were slightly higher than neat PC [94].

The fly ash waste from thermal power stations, in a finely powdered form, was chemically treated with silane and then combined with a PC matrix [95]. The tensile and flexural strengths of fly ash/PC composites were better than pure PC with the addition of fly ash at 20 wt% [95]. The introduction of talc at 10 wt% in PC improved the tensile modulus but reduced impact resistance [96].

The density of crysnanoclay/TPU/PC composites increased with an increase in crysnanoclay content (0, 5, 10, 15, and 20 wt%) due to the high density of the crysnanoclay. The tensile modulus and flexural modulus of the nanocomposites were enhanced, but elongation was reduced as a function of crysnanoclay content [97]. The tensile strength and impact strength increased by 21% and 1.9%, respectively, with the incorporation of 10 wt% crysnanoclay into the TPU/PC blend. However, when the crysnanoclay content was increased beyond 15 wt%, the tensile strength decreased. This is because the aggregation of crysnanoclay in the TPU/PC blend caused weak interaction between the crysnanoclay and the polymer blend, reducing the transfer of applied force [97].

Natural mineral fillers are inorganic materials derived from natural minerals that can reduce material costs and enhance the mechanical, physical, and thermal properties of polymer composites. Basalt fibers have mechanical properties similar to those of glass fibers, but their production cost is lower. Additionally, basalt fibers are highly thermally stable, non-toxic, and eco-friendly, making them widely used in polymer composites [24]. The addition of basalt fibers (2.5, 5, 7.5, 10, and 12.5 phr) improved the mechanical performance of PC composites, such as flexural and tensile properties, due to strong stress transfer from the matrix to the fiber. However, the impact strength decreased with increasing basalt fiber loading, as the hard basalt fibers enhanced stiffness and improved the dimensional stability of the composites [24].

The tensile modulus and density of the composites improved with an increase in titanium dioxide content, as titanium dioxide has a higher modulus and density than pure PC. This is because titanium dioxide reduces the free volume in the polymer. However, the impact strength and elongation at break decreased with higher titanium dioxide content, as cracking paths easily formed around the titanium dioxide particles, hindering the mobility of the polymer chains [25]. The optimum content of titanium dioxide was found to be 3 wt%, which exhibited the highest yield strength. The yield strength decreased when the titanium dioxide content exceeded 3 wt% [25]. F. Rouabah et al. [98] found that a quenching temperature of 40°C can improve the impact strength and elongation at break of the titanium dioxide/PC composite. This is due to the expansion of free volume, which results in increased energy absorption.

The inorganic filler calcium carbonate at a concentration of 3 wt% is commonly used as a reinforcement material in polymers [20, 21]. The inclusion of calcium carbonate in PC composites results in a 12% improvement in storage modulus values compared to neat PC. This enhancement is due to the stiffening effect of the calcium carbonate and the strong interaction between the nanofiller and the matrix, which helps to immobilize the polymer chains [20].

Dolomite is used as a filler in ABS/PC composites to reduce costs and improve mechanical properties. The addition of dolomite at concentrations of 5, 10, and 15 wt% decreases the tensile strength due to poor dispersion of the filler in the blend, creating in the composite [99]. The elongation at break and impact strength also decreases sharply with increased dolomite loading. The inclusion of 5 wt% dolomite in the blend leads to a reduction of 78% and 79% in elongation at break and impact strength, respectively. This is due to the stress concentration on the filler in the matrix, leading to the formation of cracks around dolomite and voids. However, the tensile modulus and flexural modulus of dolomite/ABS/PC composites increase with an increase in dolomite concentration due to the inherent stiffness of dolomite. The addition of 5 wt% dolomite enhances the tensile modulus by approximately 6% [99].

The increase in mica content in ABS/PC composites tends to reduce the tensile strength, elongation at break, flexural strength, and impact strength of the materials. This is due to the agglomeration of the filler and weak bonding between the mica and the polymer blend. This phenomenon creates gaps between the filler and matrix and results in poor load transfer. However, surface treatment of mica using a silane coupling agent can enhance the filler-matrix interfacial interaction, which can improve the mechanical performance of the material. The tensile modulus and flexural modulus of the composites are improved with an increase in mica content, as the high stiffness mica filler can increase the stiffness of the composites [63].

5. Thermal Properties of PC Composites

The thermal properties of materials describe their response to heating or cooling, providing insights into their relative physical characteristics at different temperatures. This information is crucial for designing and processing polymers, as well as predicting their lifespan under various conditions. Consequently, the plastics industry routinely assesses the thermal properties of polymers and polymer composites for research and quality control. Commonly examined properties include glass transition temperature (Tg), melting temperature (Tm), crystallization temperature (Tc), thermal conductivity, thermal stability, thermal expansion, and thermal shrinkage. These properties are typically investigated using techniques such as differential scanning calorimetry (DSC), thermogravimetric analysis (TGA), and thermomechanical analysis (TMA) [100].

5.1. Carbon-Based Reinforcement

Carbon-based materials such as carbon powder, carbon black, carbon fiber, carbon nanotubes, graphite powder, graphite flakes, and graphene are highly favored synthetic additives for polycarbonate due to their exceptional thermal stability and conductivity [32, 33].

Carbon fiber/PC composites exhibited superior thermal stability compared to pure PC, with this property further enhanced by increasing carbon fiber content. The improved thermal stability is attributed to the inherent thermal resistance of carbon fiber and its ability to act as a barrier, hindering the decomposition process during thermal degradation [2, 68, 69, 71]. The thermal stability of carbon nanofiber/PC composites was found to be enhanced by 54°C over neat PC [71]. Y. K. Choi et al. [68] reported that the Tg of the PC composite did not change significantly with the addition of carbon fiber into the PC matrix. Meanwhile S. Hazer et al. [69] found that the addition of carbon fiber in PLA/PC blends slightly decreased the Tg and Tm values of the composite. However, S. Kumar et al. [71] found that the incorporation of carbon nanofiber shifted the Tg of the PC composite to a higher temperature, around 10°C development, due to strong compatibility between carbon nanofiber and PC [71]. Carbon fiber/PC composites exhibited higher thermal conductivity than pure PC, with this property increasing proportionally to the carbon fiber content. This enhancement is attributed to the significantly higher thermal conductivity of carbon fiber compared to PC [101].

The addition of a small 0.5 wt% loading of CNTs to PC can enhance the thermal stability of the CNT/PC composite due to the CNTs' excellent thermal properties and their uniform dispersion within the PC matrix [73]. Furthermore, incorporating 0.5 wt% CNTs into carbon fiber/PC composites significantly improves heat transfer and dissipation, creating enhanced thermal conductivity pathways compared to pure PC. This improvement is attributed to the strong interaction between the PC matrix and the CNTs [73, 102].

The incorporation of MWCNTs can enhance the thermal stability of PC composites [78, 80, 82]. MWCNT/EMA/PC composites exhibit improved thermal stability with increasing MWCNT loading, with a 10 phr MWCNT concentration shifting the decomposition temperature approximately 20°C higher [78]. This enhancement is attributed to the strong barrier effect of MWCNTs, which inhibits the release of volatile compounds during decomposition, reducing degradation rate and delaying onset temperature [78, 80]. Additionally, small amounts of MWCNTs can increase the Tg of PC composites by immobilizing polymer chains [78, 82]. However, higher MWCNT loadings (around 3 wt%) may reduce Tg due to increased free volume caused by poor interfacial adhesion, promoting polymer chain mobility [4, 43, 76, 77, 78]. MWCNT/PC composite thermal conductivity increases moderately with MWCNT concentration due to the formation of a conductive network facilitating heat dissipation [6, 75, 80]. Optimal thermal conductivity is achieved at 20 wt% MWCNT, but further increases to 30 wt% reduce conductivity due to poor interfacial interaction, hindering heat transfer [6, 75]. Longer and larger MWCNTs offer increased surface area, enhancing thermal conductivity through phonon and electron conduction [75].

The thermal properties of hybrid nanofillers composed of graphene and MWCNT reinforced in a PC/EMA blend (5/95 wt%) were investigated [82]. Incorporating 10 wt% graphene and 10 wt% MWCNT into PC resulted in significantly enhanced thermal stability compared to the neat PC/EMA blend, with the addition of MWCNT having a particularly pronounced effect. The inclusion of these hybrid nanofillers increased the Tg of composite by 10-14°C relative to the unfilled blend. Higher concentrations of MWCNT within the hybrid nanofiller composites can further elevate thermal decomposition and Tg temperatures due to the superior thermal conductivity and barrier properties of graphene and MWCNT against volatile decomposition products [82].

The glass transition temperature (Tg) of the PC composite increased with increasing graphite flake loading due to enhanced filler-matrix interfacial adhesion [6]. Combining graphite flakes and MWCNTs as a hybrid filler further elevated Tg, as the reinforcing effect of MWCNTs immobilized polymer chains and strengthened the interfacial interaction [6]. Incorporating a hybrid filler system composed of 30% graphite flakes and 20% MWCNTs resulted in a significant 150% increase in thermal conductivity compared to pure PC [6].

The presence of graphene nanoplatelets within the PC matrix significantly enhances the thermal stability of PC composites as graphene nanoplatelet concentration increases [8, 9, 10, 11, 103]. Graphene nanoplatelets can delay composite degradation, shifting the thermal decomposition temperature upwards by as much as 50°C with the addition of only 0.5 wt% graphene nanoplatelets [10]. However, M.A. Sohel et al. [103] reported maximum thermal decomposition improvement of 50°C at a graphene nanoplatelet concentration of 3 wt%. This enhancement in thermal stability and char yield of PC nanocomposites is attributed to the interfacial adhesion between the highly thermally resistant graphene nanoplatelets and the polymer matrix, leading to the formation of a char network during decomposition. This char network restricts polymer chain mobility, obstructs gas flow, and hinders the diffusion of volatile gases into the PC matrix [9, 10, 11].

The Tg of PC composites exhibited a slight increase with the introduction of 1 wt% graphene nanoplatelets due to reduced polymer chain segment mobility caused by the nanoplatelets. However, Tg tended to decrease at higher graphene nanoplatelet loadings as increased free volume enhanced polymer chain mobility [9]. The Tg of graphene/PC nanocomposites was synergistically improved by 6°C with the addition of 3 wt% multilayer graphene, possibly due to strong interfacial interaction between the multilayer graphene nanoparticles and the PC matrix [103]. Conversely, M. T. Müller et al. [43] observed no change in Tg or Tc of PC composites with the incorporation of graphite nanoplate nanofillers. Furthermore, adding carbon fiber and CNTs to graphite nanoplate/PC composites gradually reduced Tg [43]. Graphite nanoplatelets demonstrated a more positive influence on Tg compared to carbon fiber and CNTs, as depicted in Figure 4 [43].

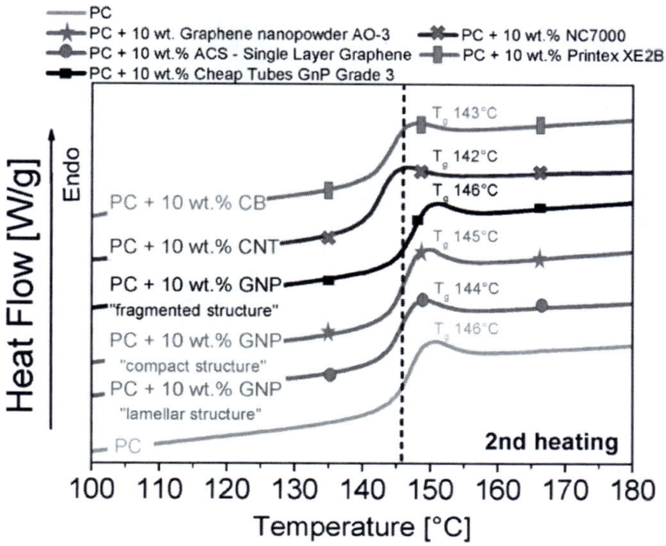

Figure 4. DSC thermogram of carbon-based fillers/PC composites. Reprinted from work of M. T.Müller et al. [43].

The thermal conductivity of graphene nanoplatelet/PC composites increased significantly with higher graphene nanoplatelet concentrations due to the high aspect ratio of graphene, which facilitated phonon transfer within the PC composites [36, 85, 109, 110]. The excellent dispersion and strong interfacial interaction between graphene nanoplatelets and PC molecules enabled phonon transport, creating a three-dimensional thermally conductive network [104, 106]. In comparison, graphite nanoplatelet/PC composites exhibited a 10% increase in thermal conductivity with a 3 vol% graphite nanoplatelet content compared to PBT/PC blends [83]. The maximum thermal conductivity of PC composites was observed with the addition of 10 wt% compact graphite nanoplatelets, resulting in a 196% increase compared to pure PC [43]. Other carbon-based fillers demonstrated only minor improvements in thermal conductivity relative to compact graphite nanoplatelets, as shown in Figure 5 [43]. The addition of CNTs could enhance the thermal conductivity of PC to a greater extent than carbon fiber [43, 84, 107].

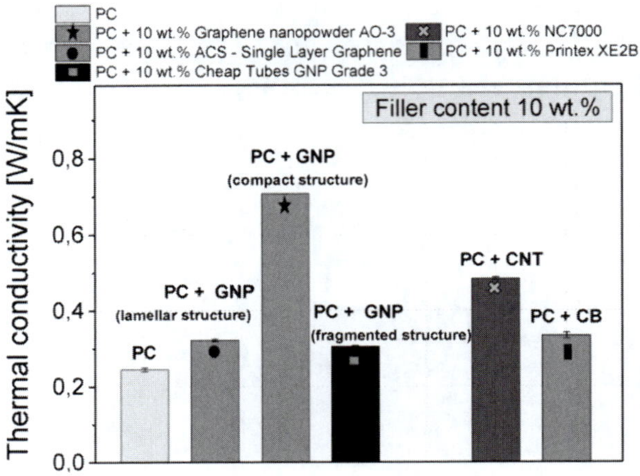

Figure 5. Thermal conductivity of carbon-based fillers/PC composites. Reprinted from work of M. T.Müller et al. [43].

Incorporating a small amount of a hybrid filler composed of 1 wt% graphene nanoplatelets and 1 wt% carbon nanotubes into boron nitride/PC composites effectively enhanced thermal conductivity by 416% and 647% compared to pure PC, respectively, and by 78% and 103% compared to boron nitride/PC composites [16]. Introducing a high thermal conductivity hybrid nanofiller consisting of 18 wt% graphene nanoplatelets (GNP) and 2 wt% multi-walled carbon nanotubes (MWCNT) significantly improved the thermal conductivity of PC composites, achieving a 479% enhancement compared to pure PC [104]. The presence of graphene flakes at a 1 wt% loading within the PC matrix enhanced the composites' thermal stability by 86°C compared to PC due to restricted polymer chain mobility [87]. Furthermore, the initial degradation temperature of PC composites increased significantly by 165°C compared to pure PC with the incorporation of 30 wt% graphene flakes, attributed to the strong interfacial interaction between graphene flakes and the PC matrix [6].

The incorporation of 3 wt% microwave-exfoliated graphite oxide into PC did not significantly enhance the thermal stability of the PC composites. Moreover, the thermal conductivity of microwave-exfoliated graphite oxide was dramatically lower than that of pristine graphene due to the presence of defects and residual functional groups on the microwave-exfoliated graphite oxide, resulting in only a minor improvement in thermal conductivity [85].

5.2. Silica-Based Materials

The research group investigated silica-based materials/PC composites due to their excellent thermal performance, high thermal stability, and thermal insulation properties. Various silica-based materials, such as fiberglass, silica particles, silica dioxide, and silica aerogel, were combined with PC to improve thermal decomposition temperature and reduce thermal conductivity.

The addition of 10 wt% glass fiber to PBT/PC shifted the Tg value to lower temperatures [89]. Incorporating nanosilica (0.1, 0.3, 0.6, and 1.0 wt%) into PC impregnated with four layers of carbon fiber (20.4 wt%) enhanced thermal stability and Tg of the composites. Optimal thermal stability was observed at 0.6 wt% nanosilica/carbon fiber/PC, with a 22°C increase in degradation temperature compared to carbon fiber/PC [67]. However, the highest Tg was achieved with 0.3 wt% nanosilica, increasing by 11°C compared to neat carbon fiber/PC due to restricted polymer chain mobility [67].

Tg of PC gradually increased with higher nano-SiO_2 content due to stronger interfacial interaction between SiO_2 and the PC matrix, hindering polymer chain mobility. Particle size of SiO_2 had a negligible effect on Tg [13]. Incorporating silica aerogel (0, 3, and 5 phr) into PMMA/PC blends reduced thermal conductivity due to silica aerogel's excellent thermal insulation and its ability to disrupt heat transfer within the composites [14].

5.3. Ceramic-Based Materials

Ceramic-based materials, known for their excellent thermal conductivity, electrical insulation, and corrosion resistance, are popular reinforcements for PC in applications such as electronic instruments, packaging, and batteries [34].

Boron nitride, a widely used ceramic material, is often combined with a PC matrix due to its excellent thermal conductivity, thermal stability, high dielectric properties, and low cost. While it can increase composite stiffness, it tends to reduce toughness [15, 16]. The thermal stability of PC composites gradually decreases with increasing boron nitride concentration due to the formation of micro-voids within the PC matrix caused by incompatibility between boron nitride particles and the PC matrix [34]. However, increasing the concentration of boron nitride functionalized with cationic polyacrylamide (CPAM) enhances both thermal stability and residue percentage of the composites. A 20 wt% loading of boron nitride functionalized with CPAM (CBN) in the PC matrix improved thermal stability by 77°C compared to pure PC due to the large surface area of CBN and its well-oriented sheets, which induced polymer chain entanglement and restricted molecular chain movement [90].

The Tg of PC composites increased with higher boron nitride loading due to restricted polymer chain mobility caused by the filler [50]. However, introducing boron nitride nanotubes decreased Tg by 4°C compared to pure PC as boron nitride nanotube content increased [91]. Tm increased with the addition of boron nitride nanotubes [91]. The incorporation of silicon carbide into PC had no significant effect on the Tm or Tg of the resulting silicon carbide/PC nanocomposites compared to pure PC [19].

The thermal conductivity of PC composites increases with increasing boron nitride loading due to interactions among boron nitride particles. The high thermal conductivity of boron nitride particles facilitates the formation of heat conductive pathways, enhancing overall composite thermal conductivity [15, 16, 34, 50, 90]. Optimal thermal conductivity in boron nitride/PC composites is achieved at a high boron nitride loading of 30 wt% [34]. However, at low boron nitride filler loadings, thermal conductivity improves slowly due to insufficient heat transfer pathways and the polymer matrix's high thermal resistance. N. Sun et al. [50], reported maximum boron nitride/PC thermal conductivity at 18.5 vol% boron nitride, attributed to increased contact area between adjacent boron nitride plates, enhancing heat conduction pathways. Similarly, X. Jin et al. [16] observed dramatic thermal conductivity improvements of 189% and 268% with 15 and 20 wt% boron nitride, respectively, compared to pure PC. Boron nitride functionalized with cationic polyacrylamide (CPAM) at 20 wt% in PC exhibited thermal conductivity 3.1 times higher than that of raw boron nitride/PC composites and pure PC, due to enhanced filler-matrix interfacial interaction [90].

The thermal stability of PC composites decreased significantly with increasing aluminum nitride content due to increased free volume and enhanced polymer chain mobility [18, 80]. A study by C. Xiao et al. [80] on hybrid filler nanocomposites of aluminum nitride nanoparticles/MWCNT reinforced in polyamide 66 (PA66)/PC blends revealed that while MWCNTs improved thermal stability, aluminum nitride reduced it. Although aluminum nitride increased energy absorption in both endothermic and exothermic DSC stages, it did not significantly improve characteristic material temperatures [18]. Conversely, the Tg of alumina/PC nanocomposites increased with alumina addition [17]. Thermal conductivity in boron nitride/alumina/PC ternary composites improved significantly with increasing alumina concentration and particle size due to enhanced filler packing density and uniform alumina distribution within the PC matrix [15].

The addition of 1 wt% titanium nitride slightly enhanced the thermal stability of PC nanocomposites. However, thermal stability decreased compared to pure PC as titanium nitride content increased [31]. The Tg of PC nanocomposites increased slightly with increasing titanium nitride content, indicating that titanium nitride did not affect characteristic temperatures but increased energy absorption in both endothermic and exothermic phases [31]. The thermal stability of silicon carbide/PC nanocomposites decreased with increasing silicon carbide loading compared to pure PC [19].

5.4. Metallic Materials

Metal materials are not commonly used as reinforcements in PC, limiting related research. X. Zhang et al. [108] found that the thermal decomposition of PC increased slightly with the addition of aluminum powder, but further increasing the aluminum powder amount did not enhance thermal stability.

5.5. Other Natural Materials

The lightweight, low-cost, and environmentally friendly nature of natural materials like cellulosic fibers from plants and agricultural waste has made them an intriguing research focus for reinforcing PC matrices, despite their inferior thermal properties. Natural mineral materials such as titanium dioxide, calcium carbonate, talc, basalt, and dolomite offer good thermal resistance, potentially improving PC's thermal properties while reducing costs. However, due to their low thermal conductivity, mineral fillers and fibers have attracted less research interest for reinforcing PC matrices.

The thermal stability of PC composites significantly improves with increasing cellulose nanocrystal content (up to a maximum studied weight fraction of 5%) due to the good dispersion of cellulose nanocrystals in the matrix, enhancing filler-matrix interaction [22]. Crysnanoclay/TPU/PC composites exhibit better thermal stability than TPU/PC blends owing to crysnanoclay's superior thermal resistance. Ash residue increases with rising nanoclay loading. The Tg of crysnanoclay/TPU/PC composites improves with increasing crysnanoclay content [97].

Pineapple leaf fiber/PC composites exhibit lower thermal stability than neat PC due to the poor thermal resistance of pineapple leaf fiber. Silane treatment of pineapple leaf fiber at 20 wt% reinforced in PC provides enhanced thermal stability compared to untreated fiber composites [93].

Fly ash waste, a finely powdered material obtained from thermal power stations, contains various inorganic oxides. Silane chemically modifies the fly ash surface. Surface treatment of fly ash is crucial for improving the thermal stability of fly ash/PC composites by increasing mechanical interlocking of PC chains within the fly ash structure [95].

The introduction of waste silk and cotton waste into PC matrices reduces the Tg value of PC composites. However, longer waste silk fibers (2.5 mm) can improve the composite's Tg value [94]. Thermal conductivity significantly decreases with increasing wood sawdust content due to wood's excellent thermal insulation properties [23].

The incorporation of talc into a PC matrix caused a decrease in the Tg of talc/PC composites [96]. Conversely, the heat deflection temperature of basalt fiber/PC composites increased by 9°C without significantly affecting the composite's Tg [24].

Titanium dioxide particles are commonly used as pigments in polymeric materials. Incorporating 0.1 wt% titanium dioxide nanowires significantly reduced the thermal stability of PC nanocomposites. However, adding 0.5 wt% titanium dioxide to PC slightly improved thermal stability compared to pure PC [58]. While titanium dioxide typically enhances thermal decomposition, the observed deterioration in titanium dioxide/PC composite properties might be attributed to poor interfacial interaction and the formation of tiny gaps between titanium dioxide and the PC matrix. Although fabricated with a large surface area, titanium dioxide nanostructures in the PC matrix did not notably alter the composite's Tg, but increased its crystallinity compared to pure PC [58].

Calcium carbonate is commonly used as a filler in polymeric materials to reduce costs and improve stiffness. Nevertheless, adding 3 wt% calcium carbonate nanoparticles decreased the composite's Tg and thermal stability due to calcium carbonate agglomeration and weak adhesion between the calcium carbonate nanofiller and the PC matrix. Calcium carbonate nanoparticles accelerated the composite's decomposition [20].

The addition of 5 wt% dolomite filler to ABS/PC did not significantly affect thermal stability but resulted in the highest Tg. However, at higher dolomite loadings, decomposition temperature decreased due to poor filler dispersion, despite dolomite's high thermal stability [99]. When dolomite was added at 10 and 15 wt%, Tg decreased below that of the polymer blend due to restricted polymeric molecule mobility caused by the mineral filler [99].

Conclusion

PC is a popular polymer matrix for creating composites with applications across various fields. Both the research community and manufacturing industries exhibit keen interest in PC composites. A wide range of reinforcing agents, including carbon-based (carbon fiber, nanotubes, graphite, graphene), silica-based (fiberglass, silica particles, dioxide, aerogel), ceramic (boron nitride, alumina, aluminum nitride, silicon nitride, silicon carbide), metal, and natural materials (cellulosic fibers, mineral fillers like titanium dioxide, calcium carbonate, talc, basalt, dolomite, mica), have been incorporated into PC matrices.

Research on PC composites focuses on understanding the impact of reinforcement materials on mechanical and thermal properties. These properties are primarily influenced by the reinforcement's characteristics, its amount, dispersion within the PC matrix, and the interfacial interaction between the reinforcement and the PC.

Carbon-based materials significantly enhance PC composites' tensile, flexural, hardness, impact, and storage modulus properties. However, elongation decreases due to increased stiffness and homogeneous reinforcement dispersion. Graphite-reinforced PC exhibits superior tensile properties compared to CNT and carbon fiber reinforced PC, while CNT/PC composites outperform carbon fiber/PC composites in tensile properties. Silica-based materials improve tensile properties and storage modulus but reduce impact and elongation. Ceramic materials enhance tensile, flexural, hardness, and storage modulus properties but diminish impact resistance and elongation. Cellulosic-based materials decrease tensile strength, elongation, and impact resistance while increasing tensile modulus. High ductility cellulosic fibers, like silk, can improve composite toughness. Mineral fillers enhance stiffness, tensile strength, modulus, flexural strength, modulus, and storage modulus but reduce ductility.

Carbon-based materials improve overall thermal properties, including thermal stability, Tg, and thermal conductivity. Graphene/PC composites exhibit higher Tg compared to CNT/PC and carbon fiber/PC composites. Silica-based materials enhance thermal stability and Tg but reduce thermal conductivity relative to pure PC. Ceramic fillers can increase decomposition temperature, Tg, and thermal conductivity, while cellulosic-based reinforcements generally decrease thermal properties. Mineral filler incorporation has a minimal impact on the thermal properties of PC composites.

Interfacial interaction between the reinforcement and the polymer matrix is crucial for developing optimal mechanical and thermal properties in polymer composites, as evidenced by various studies. This interaction facilitates stress transfer from the PC matrix to the reinforcement and reduces voids at the reinforcement-matrix interface. Surface treatments such as silane, alkali, coupling agent, ultrasonic, isocyanate, mercerization, steam explosion, peroxide, and enzymatic treatments can enhance reinforcement-matrix interaction [109, 110, 111, 112, 113, 114].

References

[1] Kyriacos, D., "Chapter 17 - Polycarbonates," in *Brydson's Plastics Materials (Eighth Edition)*, M. Gilbert, Ed., Butterworth-Heinemann, 2017, pp. 457–485. doi: 10.1016/B978-0-323-35824-8.00017-7.

[2] Ozkan, C., N. Gamze Karsli, A. Aytac, and V. Deniz, "Short carbon fiber reinforced polycarbonate composites: Effects of different sizing materials," *Compos. Part B Eng.*, vol. 62, pp. 230–235, Jun. 2014, doi: 10.1016/j.compositesb.2014.03.002.

[3] Monti, M., M. Palenzona, F. Fiorino, F. Baudach, A. Onnis, and A. Romeo, "Design, manufacturing and FEA prediction of the mechanical behavior of a hybrid-molded polycarbonate/continuous carbon fiber reinforced composite component," *Compos. Part B Eng.*, vol. 238, p. 109891, Jun. 2022, doi: 10.1016/j.compositesb.2022.109891.

[4] Sung, Y. T., C. K. Kum, H. S. Lee, N. S. Byon, H. G. Yoon, and W. N. Kim, "Dynamic mechanical and morphological properties of polycarbonate/multi-walled carbon nanotube composites," *Polymer*, vol. 46, no. 15, pp. 5656–5661, Jul. 2005, doi: 10.1016/j.polymer.2005.04.075.

[5] "Distribution of Carbon Nanotubes in Polycarbonate-Based Blends for Electromagnetic Interference Shielding | ACS Applied Nano Materials." Accessed: Jan. 31, 2023. [Online]. Available: https://pubs.acs.org/doi/full/10.1021/acsanm.1c03442?casa_token=3bBXOkcqJokAAAAA%3Ak8gjfdhAgH97EHYk3SYFZHFlh3rmh2W4JCEPJ6tQJ9gpR5_tcKcwshgxRP13NiWdZ8jnasplPMv00WjT3g.

[6] Pradhan, S. S., L. Unnikrishnan, S. Mohanty, and S. K. Nayak, "Effect of graphite flake and multi-walled carbon nanotube on thermal, mechanical, electrical, and electromagnetic interference shielding properties of polycarbonate nanocomposite," *Polym. Compos.*, vol. 42, no. 8, pp. 4043–4055, 2021, doi: 10.1002/pc.26115.

[7] Kim, H. and C. W. Macosko, "Processing-property relationships of polycarbonate/graphene composites," *Polymer*, vol. 50, no. 15, pp. 3797–3809, Jul. 2009, doi: 10.1016/j.polymer.2009.05.038.

[8] Heidar Pour, R., M. Soheilmoghaddam, A. Hassan, and S. Bourbigot, "Flammability and thermal properties of polycarbonate/acrylonitrile-butadiene-styrene nanocomposites reinforced with multilayer graphene," *Polym. Degrad. Stab.*, vol. 120, pp. 88–97, Oct. 2015, doi: 10.1016/j.polymdegradstab.2015.06.013.

[9] Wijerathne, D., Y. Gong, S. Afroj, N. Karim, and C. Abeykoon, "Mechanical and thermal properties of graphene nanoplatelets-reinforced recycled polycarbonate composites," *Int. J. Lightweight Mater. Manuf.*, vol. 6, no. 1, pp. 117–128, Mar. 2023, doi: 10.1016/j.ijlmm.2022.09.001.

[10] Sabet, M. and E. Mohammadian, "The inclusion of graphene nanoplatelet on the mechanical, thermal, and electrical characteristics of polycarbonate," *Polym. Bull.*, vol. 80, no. 2, pp. 2153–2169, Feb. 2023, doi: 10.1007/s00289-022-04156-8.

[11] Pour, R. H., A. Hassan, M. Soheilmoghaddam, and H. C. Bidsorkhi, "Mechanical, thermal, and morphological properties of graphene reinforced polycarbonate/acrylonitrile butadiene styrene nanocomposites," *Polym. Compos.*, vol. 37, no. 6, pp. 1633–1640, 2016, doi: 10.1002/pc.23335.

[12] Lan, Z. et al., "Color Changes and Mechanical Properties of Glass Fiber Reinforced Polycarbonate Composites after Thermal Aging," *Polymers*, vol. 14, no. 2, Art. no. 2, Jan. 2022, doi: 10.3390/polym14020222.

[13] Nomai, J. and A. K. Schlarb, "Effects of nanoparticle size and concentration on optical, toughness, and thermal properties of polycarbonate," *J. Appl. Polym. Sci.*, vol. 136, no. 23, p. 47634, 2019, doi: 10.1002/app.47634.

[14] Estaji, S., A. Paydayesh, S. R. Mousavi, H. A. Khonakdar, and M. M. Abiyati, "Polycarbonate/poly(methyl methacrylate)/silica aerogel blend composites for advanced transparent thermal insulations: Mechanical, thermal, and optical studies," *Polym. Compos.*, vol. 42, no. 10, pp. 5323–5334, 2021, doi: 10.1002/pc.26226.

[15] Zhou, C., Y. Bai, H. Zou, and S. Zhou, "Improving Thermal Conductivity of Injection Molded Polycarbonate/Boron Nitride Composites by Incorporating Spherical Alumina Particles: The Influence of Alumina Particle Size," *Polymers*, vol. 14, no. 17, Art. no. 17, Jan. 2022, doi: 10.3390/polym14173477.

[16] Jin, X., J. Wang, L. Dai, W. Wang, and H. Wu, "Largely enhanced thermal conductive, dielectric, mechanical and anti-dripping performance in polycarbonate/boron nitride composites with graphene nanoplatelet and carbon nanotube," *Compos. Sci. Technol.*, vol. 184, p. 107862, Nov. 2019, doi: 10.1016/j.compscitech.2019.107862.

[17] Aghajani Derazkola, H. and A. Simchi, "Processing and characterizations of polycarbonate/alumina nanocomposites by additive powder fed friction stir processing," *Thin-Walled Struct.*, vol. 157, p. 107086, Dec. 2020, doi: 10.1016/j.tws.2020.107086.

[18] Vidakis, N. et al., "Thermomechanical Response of Polycarbonate/Aluminum Nitride Nanocomposites in Material Extrusion Additive Manufacturing," *Materials*, vol. 15, no. 24, Art. no. 24, Jan. 2022, doi: 10.3390/ma15248806.

[19] Petousis, M. et al., "Silicon Carbide Nanoparticles as a Mechanical Boosting Agent in Material Extrusion 3D-Printed Polycarbonate," *Polymers*, vol. 14, no. 17, Art. no. 17, Jan. 2022, doi: 10.3390/polym14173492.

[20] Charde, S. J., S. S. Sonawane, S. H. Sonawane, and S. Navin, "Influence of functionalized calcium carbonate nanofillers on the properties of melt-extruded polycarbonate composites," *Chem. Eng. Commun.*, vol. 205, no. 4, pp. 492–505, Apr. 2018, doi: 10.1080/00986445.2017.1404459.

[21] Techawinyutham, L., N. Sumrith, R. Srisuk, W. Techawinyutham, S. Siengchin, and S. Mavinkere Rangappa, "Thermo-mechanical, rheological and morphology properties of polypropylene composites: Residual CaCO3 as a sustainable by-product," *Polym. Compos.*, vol. 42, no. 9, pp. 4643–4659, 2021, doi: 10.1002/pc.26175.

[22] Xu, W. et al., "Cellulose nanocrystals as organic nanofillers for transparent polycarbonate films," *J. Nanoparticle Res.*, vol. 15, no. 4, p. 1562, Mar. 2013, doi: 10.1007/s11051-013-1562-0.

[23] Wimonsong, W., P. Threepopnatkul, and C. Kulsetthanchalee, "Thermal Conductivity and Mechanical Properties of Wood Sawdust/Polycarbonate Composites," *Mater. Sci. Forum*, vol. 714, pp. 139–146, 2012, doi: 10.4028/www.scientific.net/MSF.714.139.

[24] Jang, K.-S., "Mechanics and rheology of basalt fiber-reinforced polycarbonate composites," *Polymer*, vol. 147, pp. 133–141, Jul. 2018, doi: 10.1016/j.polymer.2018.06.004.

[25] Rouabah, F. et al., "Mechanical and thermal properties of polycarbonate. II. Influence of titanium dioxide content and quenching on pigmented polycarbonate," *J. Appl. Polym. Sci.*, vol. 106, no. 4, pp. 2710–2717, 2007, doi: 10.1002/app.26807.

[26] Aguirre-Vargas, F., "Chapter 11 - Thermoset coatings," in *Thermosets (Second Edition)*, Q. Guo, Ed., Elsevier, 2018, pp. 369–400. doi: 10.1016/B978-0-08-101021-1.00011-3.

[27] Nejatian, T., S. Pezeshki, and A. U. Yaqin Syed, "5 - Acrylic denture base materials," in *Advanced Dental Biomaterials*, Z. Khurshid, S. Najeeb, M. S. Zafar, and F. Sefat, Eds., Woodhead Publishing, 2019, pp. 79–104. doi: 10.1016/B978-0-08-102476-8.00005-0.

[28] Begum, S. A., A. V. Rane, and K. Kanny, "Chapter 20 - Applications of compatibilized polymer blends in automobile industry," in *Compatibilization of Polymer Blends*, A. A. r. and S. Thomas, Eds., Elsevier, 2020, pp. 563–593. doi: 10.1016/B978-0-12-816006-0.00020-7.

[29] Amir, S. M. M. et al., "16 - Nondestructive testing method for Kevlar and natural fiber and their hybrid composites," in *Durability and Life Prediction in Biocomposites, Fibre-Reinforced Composites and Hybrid Composites*, M. Jawaid, M. Thariq, and N. Saba, Eds., in Woodhead Publishing Series in Composites Science and Engineering. Woodhead Publishing, 2019, pp. 367–388. doi: 10.1016/B978-0-08-102290-0.00016-7.

[30] Afzal, A. and Y. Nawab, "5 - Polymer composites," in *Composite Solutions for Ballistics*, Y. Nawab, S. M. Sapuan, and K. Shaker, Eds., in Woodhead Publishing Series in Composites Science and Engineering. Woodhead Publishing, 2021, pp. 139–152. doi: 10.1016/B978-0-12-821984-3.00003-6.

[31] Vidakis, N. et al., "On the thermal and mechanical performance of Polycarbonate / Titanium Nitride nanocomposites in material extrusion additive manufacturing," *Compos. Part C Open Access*, vol. 8, p. 100291, Jul. 2022, doi: 10.1016/j.jcomc.2022.100291.

[32] Harussani, M. M., S. M. Sapuan, G. Nadeem, T. Rafin, and W. Kirubaanand, "Recent applications of carbon-based composites in defence industry: A review," *Def. Technol.*, vol. 18, no. 8, pp. 1281–1300, Aug. 2022, doi: 10.1016/j.dt.2022.03.006.

[33] Barbir, F., "CHAPTER 4 - Main Cell Components, Materials Properties and Processes," in *PEM Fuel Cells*, F. Barbir, Ed., Burlington: Academic Press, 2005, pp. 73–113. doi: 10.1016/B978-012078142-3/50005-7.

[34] Zhou, S., Y. Shi, Y. Bai, M. Liang, and H. Zou, "Preparation of thermally conductive polycarbonate/boron nitride composites with balanced mechanical properties," *Polym. Compos.*, vol. 41, no. 12, pp. 5418–5427, 2020, doi: 10.1002/pc.25805.

[35] Soutis, C., "Carbon fiber reinforced plastics in aircraft construction," *Mater. Sci. Eng. A*, vol. 412, no. 1, pp. 171–176, Dec. 2005, doi: 10.1016/j.msea.2005.08.064.

[36] Eom, W. et al., "Carbon nanotube-reduced graphene oxide fiber with high torsional strength from rheological hierarchy control," *Nat. Commun.*, vol. 12, no. 1, Art. no. 1, Jan. 2021, doi: 10.1038/s41467-020-20518-0.

[37] Huang, X., "Fabrication and Properties of Carbon Fibers," *Materials*, vol. 2, no. 4, Art. no. 4, Dec. 2009, doi: 10.3390/ma2042369.

[38] Newcomb, B. A., "Processing, structure, and properties of carbon fibers," *Compos. Part Appl. Sci. Manuf.*, vol. 91, pp. 262–282, Dec. 2016, doi: 10.1016/j.compositesa.2016.10.018.

[39] Harris, P. J. F., "Carbon nanotube composites," *Int. Mater. Rev.*, vol. 49, no. 1, pp. 31–43, Feb. 2004, doi: 10.1179/095066004225010505.

[40] Liu, Y., C. Nie, X. Liu, X. Xu, Z. Sun, and L. Pan, "Review on carbon-based composite materials for capacitive deionization," *RSC Adv.*, vol. 5, no. 20, pp. 15205–15225, Jan. 2015, doi: 10.1039/C4RA14447C.

[41] Das, P., S. Banerjee, and N. C. Das, "23 - Polymer-graphene composite in aerospace engineering," in *Polymer Nanocomposites Containing Graphene*, M. Rahaman, L. Nayak, I. A. Hussein, and N. C. Das, Eds., in Woodhead Publishing Series in Composites Science and Engineering. , Woodhead Publishing, 2022, pp. 683–711. doi: 10.1016/B978-0-12-821639-2.00001-X.

[42] Bhattacharjee, Y., S. Biswas, and S. Bose, "Chapter 5 - Thermoplastic polymer composites for EMI shielding applications," in *Materials for Potential EMI Shielding Applications*, K. Joseph, R. Wilson, and G. George, Eds., Elsevier, 2020, pp. 73–99. doi: 10.1016/B978-0-12-817590-3.00005-1.

[43] Müller, M. T., K. Hilarius, M. Liebscher, D. Lellinger, I. Alig, and P. Pötschke, "Effect of Graphite Nanoplate Morphology on the Dispersion and Physical Properties of Polycarbonate Based Composites," *Materials*, vol. 10, no. 5, Art. no. 5, May 2017, doi: 10.3390/ma10050545.

[44] Phiri, J., P. Gane, and T. C. Maloney, "General overview of graphene: Production, properties and application in polymer composites," *Mater. Sci. Eng. B*, vol. 215, pp. 9–28, Jan. 2017, doi: 10.1016/j.mseb.2016.10.004.

[45] Hashmi, A. et al., "Potentialities of graphene and its allied derivatives to combat against SARS-CoV-2 infection," *Mater. Today Adv.*, vol. 13, p. 100208, Mar. 2022, doi: 10.1016/j.mtadv.2022.100208.

[46] Ciriminna, R., A. Fidalgo, V. Pandarus, F. Béland, L. M. Ilharco, and M. Pagliaro, "The Sol–Gel Route to Advanced Silica-Based Materials and Recent Applications," *Chem. Rev.*, vol. 113, no. 8, pp. 6592–6620, Aug. 2013, doi: 10.1021/cr300399c.

[47] Wallenberger, F. T., "Structural Silicate and Silica Glass Fibers," in *Advanced Inorganic Fibers: Process - Structure - Properties - Applications*, F. T. Wallenberger, R. Naslain, J. B. Macchesney, H. D. Ackler, and F. T. Wallenberger, Eds., in Materials Technology Series. , Boston, MA: Springer US, 2000, pp. 129–168. doi: 10.1007/978-1-4419-8722-8_6.

[48] Sathishkumar, T., S. Satheeshkumar, and J. Naveen, "Glass fiber-reinforced polymer composites – a review," *J. Reinf. Plast. Compos.*, vol. 33, no. 13, pp. 1258–1275, Jul. 2014, doi: 10.1177/0731684414530790.

[49] Simonova, L. G. et al., "Catalysts Based on Fiberglass Supports: I. Physicochemical Properties of Silica Fiberglass Supports," *Kinet. Catal.*, vol. 42, no. 5, pp. 693–702, Sep. 2001, doi: 10.1023/A:1012331916839.

[50] Sun, N. et al., "Hot-pressing induced orientation of boron nitride in polycarbonate composites with enhanced thermal conductivity," *Compos. Part Appl. Sci. Manuf.*, vol. 110, pp. 45–52, Jul. 2018, doi: 10.1016/j.compositesa.2018.04.010.

[51] Ortiz, A. L., F. Sánchez-Bajo, F. L. Cumbrera, and F. Guiberteau, "X-ray powder diffraction analysis of a silicon carbide-based ceramic," *Mater. Lett.*, vol. 49, no. 2, pp. 137–145, Jun. 2001, doi: 10.1016/S0167-577X(00)00358-X.

[52] Abdulkareem, S. A., M. K. Amosa, A. G. Adeniyi, M. M. Magaji, and R. A. Ajibola, "Effect of metallic fillers on the hardness of polystyrene composites: An experimental investigation," *IOP Conf. Ser. Mater. Sci. Eng.*, vol. 640, no. 1, p. 012058, 2019, doi: 10.1088/1757-899X/640/1/012058.

[53] Tavman, I. H., "Thermal and mechanical properties of aluminum powder-filled high-density polyethylene composites," *J. Appl. Polym. Sci.*, vol. 62, no. 12, pp. 2161–2167, 1996, doi: 10.1002/(SICI)1097-4628(19961219)62:12<2161::AID-APP19>3.0.CO;2-8.

[54] Videira-Quintela, D., O. Martin, and G. Montalvo, "Recent advances in polymer-metallic composites for food packaging applications," *Trends Food Sci. Technol.*, vol. 109, pp. 230–244, Mar. 2021, doi: 10.1016/j.tifs.2021.01.020.

[55] Haris, N. I. N. et al., "Dynamic mechanical properties of natural fiber reinforced hybrid polymer composites: a review," *J. Mater. Res. Technol.*, vol. 19, pp. 167–182, Jul. 2022, doi: 10.1016/j.jmrt.2022.04.155.

[56] Ahmad Fauzi, A. A., A. F. Osman, A. A. Alrashdi, Z. Mustafa, and K. A. Abdul Halim, "On the Use of Dolomite as a Mineral Filler and Co-Filler in the Field of Polymer Composites: A Review," *Polymers*, vol. 14, no. 14, Art. no. 14, Jan. 2022, doi: 10.3390/polym14142843.

[57] Tao, H., Y. He, and X. Zhao, "Preparation and characterization of calcium carbonate–titanium dioxide core–shell (CaCO3@TiO2) nanoparticles and application in the papermaking industry," *Powder Technol.*, vol. 283, pp. 308–314, Oct. 2015, doi: 10.1016/j.powtec.2015.05.039.

[58] Khalid, Y., A. Achour, M. A. Akram, and M. Islam, "Polycarbonate/Titania Composites Incorporating TiO2 with Different Nanoscale Morphologies for Enhanced Environmental Stress Cracking Resistance in Dioctyl Phthalate," *Polymers*, vol. 14, no. 17, Art. no. 17, Jan. 2022, doi: 10.3390/polym14173693.

[59] Techawinyutham, L., A. Frick, and S. Siengchin, "Polypropylene/Maleic Anhydride Grafted Polypropylene (MAgPP)/Coconut Fiber Composites," *Adv. Mech. Eng.*, vol. 8, no. 5, p. 1687814016645446, May 2016, doi: 10.1177/1687814016645446.

[60] Sumrith, N., L. Techawinyutham, M. R. Sanjay, R. Dangtungee, and S. Siengchin, "Characterization of Alkaline and Silane Treated Fibers of 'Water Hyacinth Plants' and Reinforcement of 'Water Hyacinth Fibers' with Bioepoxy to Develop Fully Biobased Sustainable Ecofriendly Composites," *J. Polym. Environ.*, Jun. 2020, doi: 10.1007/s10924-020-01810-y.

[61] Kumar, S., A. Manna, and R. Dang, "A review on applications of natural Fiber-Reinforced composites (NFRCs)," *Mater. Today Proc.*, vol. 50, pp. 1632–1636, Jan. 2022, doi: 10.1016/j.matpr.2021.09.131.

[62] Li, Z., A. Shen, G. Zeng, Z. Chen, and Y. Guo, "Research progress on properties of basalt fiber-reinforced cement concrete," *Mater. Today Commun.*, vol. 33, p. 104824, Dec. 2022, doi: 10.1016/j.mtcomm.2022.104824.

[63] Asyadi, F., M. Jawaid, A. Hassan, and M. U. Wahit, "Mechanical Properties of Mica-Filled Polycarbonate/Poly(Acrylonitrile-Butadiene-Styrene) Composites," *Polym.-Plast. Technol. Eng.*, vol. 52, no. 7, pp. 727–736, May 2013, doi: 10.1080/03602559.2012.762672.

[64] Grellmann, W. and S. Seidler, "4 - Mechanical Properties of Polymers," in *Polymer Testing (Third Edition)*, W. Grellmann and S. Seidler, Eds., Hanser, 2022, pp. 71–227. doi: 10.3139/9781569908075.004.

[65] Kara, Y., V. Acar, and M. O. Seydibeyoglu, "4 - Mechanical properties of nanoparticle-based polymer composites," in *Nanoparticle-Based Polymer Composites*, S. Mavinkere Rangappa, J. Parameswaranpillai, T. G. Yashas Gowda, S. Siengchin, and M. O. Seydibeyoglu, Eds., in Woodhead Publishing Series in Composites Science and Engineering. Woodhead Publishing, 2022, pp. 95–108. doi: 10.1016/B978-0-12-824272-8.00007-5.

[66] Ramesh, M. and C. Deepa, "Chapter Fifteen - Mechanical properties of natural and synthetic fiber reinforced hybrid composites," in *Hybrid Natural Fiber Composites*, A. Khan, S. M. Rangappa, S. Siengchin, M. Jawaid, and A. M. Asiri, Eds., in Woodhead Publishing Series in Composites Science and Engineering.

Woodhead Publishing, 2021, pp. 309–325. doi: 10.1016/B978-0-12-819900-8.00013-1.

[67] Fathalian, M., M. Ghorbanzadeh Ahangari, and A. Fereidoon, "Effect of Nanosilica on the Mechanical and Thermal Properties of Carbon Fiber/Polycarbonate Laminates," *Fibers Polym.*, vol. 20, no. 8, pp. 1684–1689, Aug. 2019, doi: 10.1007/s12221-019-1158-0.

[68] Choi, Y. K., K. I. Sugimoto, S. M. Song, and M. Endo, "Mechanical and thermal properties of vapor-grown carbon nanofiber and polycarbonate composite sheets," *Mater. Lett.*, vol. 59, no. 27, pp. 3514–3520, Nov. 2005, doi: 10.1016/j.matlet.2005.05.082.

[69] Hazer, S., M. Coban, and A. Aytac, "A study on carbon fiber reinforced poly(lactic acid)/polycarbonate composites," *J. Appl. Polym. Sci.*, vol. 135, no. 48, p. 46881, 2018, doi: 10.1002/app.46881.

[70] Lively, B., S. Kumar, L. Tian, B. Li, and W.-H. Zhong, "Mechanical, Thermal and Morphological Characterization of Polycarbonate/Oxidized Carbon Nanofiber Composites Produced with a Lean 2-Step Manufacturing Process," *J. Nanosci. Nanotechnol.*, vol. 11, no. 5, pp. 3929–3937, May 2011, doi: 10.1166/jnn.2011.3853.

[71] Kumar, S., B. Lively, L. L. Sun, B. Li, and W. H. Zhong, "Highly dispersed and electrically conductive polycarbonate/oxidized carbon nanofiber composites for electrostatic dissipation applications," *Carbon*, vol. 48, no. 13, pp. 3846–3857, Nov. 2010, doi: 10.1016/j.carbon.2010.06.050.

[72] Wu, Y., D. Dhamodharan, Z. Wang, R. Wang, and L. Wu, "Effect of electrophoretic deposition followed by solution pre-impregnated surface modified carbon fiber-carbon nanotubes on the mechanical properties of carbon fiber reinforced polycarbonate composites," *Compos. Part B Eng.*, vol. 195, p. 108093, Aug. 2020, doi: 10.1016/j.compositesb.2020.108093.

[73] Baek, Y.-M., P.-S. Shin, J.-H. Kim, H.-S. Park, K. L. DeVries, and J.-M. Park, "Thermal transfer, interfacial, and mechanical properties of carbon fiber/polycarbonate-CNT composites using infrared thermography," *Polym. Test.*, vol. 81, p. 106247, Jan. 2020, doi: 10.1016/j.polymertesting.2019.106247.

[74] Bagotia, N., V. Choudhary, and D. K. Sharma, "Studies on toughened polycarbonate/multiwalled carbon nanotubes nanocomposites," *Compos. Part B Eng.*, vol. 124, pp. 101–110, Sep. 2017, doi: 10.1016/j.compositesb.2017.05.037.

[75] Sharma, S., R. Chandra, P. Kumar, and N. Kumar, "Thermo-mechanical characterization of multi-walled carbon nanotube reinforced polycarbonate composites: A molecular dynamics approach," *Comptes Rendus Mécanique*, vol. 343, no. 5, pp. 371–396, May 2015, doi: 10.1016/j.crme.2015.03.002.

[76] Guo, J. et al., "Aspect ratio effects of multi-walled carbon nanotubes on electrical, mechanical, and thermal properties of polycarbonate/MWCNT composites," *J. Polym. Sci. Part B Polym. Phys.*, vol. 52, no. 1, pp. 73–83, 2014, doi: 10.1002/polb.23402.

[77] Gong, Z., J. Gong, X. Yan, S. Gao, and B. Wang, "Investigation of the Effects of Temperature and Strain on the Damping Properties of Polycarbonate/Multiwalled

Carbon Nanotube Composites," *J. Phys. Chem. C*, vol. 115, no. 38, pp. 18468–18472, Sep. 2011, doi: 10.1021/jp205354e.

[78] Bagotia, N. and D. K. Sharma, "Systematic study of dynamic mechanical and thermal properties of multiwalled carbon nanotube reinforced polycarbonate/ethylene methyl acrylate nanocomposites," *Polym. Test.*, vol. 73, pp. 425–432, Feb. 2019, doi: 10.1016/j.polymertesting.2018.12.006.

[79] Kim, K. H. and W. H. Jo, "A strategy for enhancement of mechanical and electrical properties of polycarbonate/multi-walled carbon nanotube composites," *Carbon*, vol. 47, no. 4, pp. 1126–1134, Apr. 2009, doi: 10.1016/j.carbon.2008.12.043.

[80] Xiao, C., X. Leng, X. Zhang, K. Zheng, and X. Tian, "Improved thermal properties by controlling selective distribution of AlN and MWCNT in immiscible polycarbonate (PC)/Polyamide 66 (PA66) composites," *Compos. Part Appl. Sci. Manuf.*, vol. 110, pp. 133–141, Jul. 2018, doi: 10.1016/j.compositesa.2018.03.030.

[81] Cho, B.-G., S.-H. Hwang, M. Park, J. K. Park, Y.-B. Park, and H. G. Chae, "The effects of plasma surface treatment on the mechanical properties of polycarbonate/carbon nanotube/carbon fiber composites," *Compos. Part B Eng.*, vol. 160, pp. 436–445, Mar. 2019, doi: 10.1016/j.compositesb.2018.12.062.

[82] Bagotia, N., V. Choudhary, and D. K. Sharma, "Synergistic effect of graphene/multiwalled carbon nanotube hybrid fillers on mechanical, electrical and EMI shielding properties of polycarbonate/ethylene methyl acrylate nanocomposites," *Compos. Part B Eng.*, vol. 159, pp. 378–388, Feb. 2019, doi: 10.1016/j.compositesb.2018.10.009.

[83] Wen, B. and X. Zheng, "Effect of the selective distribution of graphite nanoplatelets on the electrical and thermal conductivities of a polybutylene terephthalate/polycarbonate blend," *Compos. Sci. Technol.*, vol. 174, pp. 68–75, Apr. 2019, doi: 10.1016/j.compscitech.2019.02.017.

[84] King, J. A. et al., "Characterization of exfoliated graphite nanoplatelets/polycarbonate composites: electrical and thermal conductivity, and tensile, flexural, and rheological properties," *J. Compos. Mater.*, vol. 46, no. 9, pp. 1029–1039, May 2012, doi: 10.1177/0021998311414073.

[85] Potts, J. R., S. Murali, Y. Zhu, X. Zhao, and R. S. Ruoff, "Microwave-Exfoliated Graphite Oxide/Polycarbonate Composites," *Macromolecules*, vol. 44, no. 16, pp. 6488–6495, Aug. 2011, doi: 10.1021/ma2007317.

[86] Zakaulla, M., F. Parveen, Amreen, Harish, and N. Ahmad, "Artificial neural network based prediction on tribological properties of polycarbonate composites reinforced with graphene and boron carbide particle," *Mater. Today Proc.*, vol. 26, pp. 296–304, Jan. 2020, doi: 10.1016/j.matpr.2019.11.276.

[87] E. Lago, E., P. S. Toth, G. Pugliese, V. Pellegrini, and F. Bonaccorso, "Solution blending preparation of polycarbonate/graphene composite: boosting the mechanical and electrical properties," *RSC Adv.*, vol. 6, no. 100, pp. 97931–97940, Oct. 2016, doi: 10.1039/C6RA21962D.

[88] Manjunath, A., "Effect of Short Glass Fiber Reinforcement on Characteristic of Polymer Matrix (Polycarbonate) An Experimental Study," *Int. Conf. Adv. ...*, Jan. 2007, Accessed: Mar. 03, 2023. [Online]. Available: https://www.academia.edu/

3382968/Effect_of_Short_Glass_Fiber_Reinforcement_on_Characteristic_of_Polymer_Matrix_Polycarbonate_an_Experimental_Study.

[89] Kuram, E., B. Ozcelik, and F. Yilmaz, "The influence of recycling number on the mechanical, chemical, thermal and rheological properties of poly(butylene terephthalate)/polycarbonate binary blend and glass-fibre-reinforced composite," *J. Thermoplast. Compos. Mater.*, vol. 29, no. 10, pp. 1443–1457, Oct. 2016, doi: 10.1177/0892705715569823.

[90] Wang, J. et al., "Noncovalent functionalization of boron nitride and its effect on the thermal conductivity of polycarbonate composites," *J. Appl. Polym. Sci.*, vol. 134, no. 25, 2017, doi: 10.1002/app.44978.

[91] Guan, J. et al., "Boron nitride nanotubes reinforced polycarbonate nanocomposites," *Mater. Today Commun.*, vol. 20, p. 100586, Sep. 2019, doi: 10.1016/j.mtcomm.2019.100586.

[92] Shemelya, C. M. et al., "Mechanical, Electromagnetic, and X-ray Shielding Characterization of a 3D Printable Tungsten–Polycarbonate Polymer Matrix Composite for Space-Based Applications," *J. Electron. Mater.*, vol. 44, no. 8, pp. 2598–2607, Aug. 2015, doi: 10.1007/s11664-015-3687-7.

[93] Threepopnatkul, P., N. Kaerkitcha, and N. Athipongarporn, "Effect of surface treatment on performance of pineapple leaf fiber–polycarbonate composites," *Compos. Part B Eng.*, vol. 40, no. 7, pp. 628–632, Oct. 2009, doi: 10.1016/j.compositesb.2009.04.008.

[94] Taşdemır, M., D. Koçak, İ. Usta, M. Akalin, and N. Merdan, "Properties of Recycled Polycarbonate/Waste Silk and Cotton Fiber Polymer Composites," *Int. J. Polym. Mater. Polym. Biomater.*, vol. 57, no. 8, pp. 797–805, Jun. 2008, doi: 10.1080/00914030802089138.

[95] Parvaiz, M. R., "Influence of silane-coupling agents on the performance of morphological, mechanical, thermal, electrical, and rheological properties of polycarbonate/fly ash composites," *Polym. Compos.*, vol. 33, no. 10, pp. 1798–1808, 2012, doi: 10.1002/pc.22325.

[96] Sung, Y. T., P. D. Fasulo, W. R. Rodgers, Y. T. Yoo, Y. Yoo, and D. R. Paul, "Properties of polycarbonate/acrylonitrile-butadiene-styrene/talc composites," *J. Appl. Polym. Sci.*, vol. 124, no. 2, pp. 1020–1030, 2012, doi: 10.1002/app.35147.

[97] Pashaei, S., Siddaramaiah, A. A. Syed, G. R. Reddy, and S. Srinivasa, "The Influence of Crysnanoclay Addition on Mechanical and Thermal Properties of Thermoplastic Polyurethane Elastomeric with Polycarbonate Nanocomposites," *Polym.-Plast. Technol. Eng.*, vol. 51, no. 9, pp. 911–919, Jun. 2012, doi: 10.1080/03602559.2012.671426.

[98] Rouabah, F., D. Dadache, M. Fois, and N. Haddaoui, "Effect of the quenching temperature on the mechanical and thermophysical properties of polycarbonate pigmented with titanium dioxide," *J. Polym. Eng.*, vol. 34, no. 7, pp. 657–663, Sep. 2014, doi: 10.1515/polyeng-2014-0021.

[99] Hassan, A., O. E. Ezenkwa, A. S. Ismail, M. A. A. Saidi, R. A. Rashid, and R. Arjmandi, "Mechanical and Thermal Properties of Dolomite Filled Polycarbonate/Acrylonitrile Butadiene Styrene Composites," *Perintis Ejournal*, vol. 9, no. 1, Art. no. 1, 2019.

[100] Menczel, J. D. and R. B. Prime, Eds., *Thermal Analysis of Polymers: Fundamentals and Applications*, 1st edition. Hoboken, N.J: Wiley, 2009.

[101] Nithikarnjanatharn, J., H. Ueda, S. Tanoue, H. Uematsu, and Y. Iemoto, "The rheological behavior and thermal conductivity of melt-compounded polycarbonate/vapor-grown carbon fiber composites," *Polym. J.*, vol. 44, no. 5, Art. no. 5, May 2012, doi: 10.1038/pj.2011.149.

[102] Ajayan, P. M. and J. M. Tour, "Nanotube composites," *Nature*, vol. 447, no. 7148, Art. no. 7148, Jun. 2007, doi: 10.1038/4471066a.

[103] Sohel, M. A., A. Mondal, P. M. Arif, S. Thomas, and A. SenGupta, "Effects of graphene on thermal properties and thermal stability of polycarbonate/graphene nanocomposite," *J. Thermoplast. Compos. Mater.*, vol. 36, no. 1, pp. 326–344, Jan. 2023, doi: 10.1177/0892705721997887.

[104] Yu, J., H. K. Choi, H. S. Kim, and S. Y. Kim, "Synergistic effect of hybrid graphene nanoplatelet and multi-walled carbon nanotube fillers on the thermal conductivity of polymer composites and theoretical modeling of the synergistic effect," *Compos. Part Appl. Sci. Manuf.*, vol. 88, pp. 79–85, Sep. 2016, doi: 10.1016/j.compositesa.2016.05.022.

[105] Ren, L., K. Pashayi, H. R. Fard, S. P. Kotha, T. Borca-Tasciuc, and R. Ozisik, "Engineering the coefficient of thermal expansion and thermal conductivity of polymers filled with high aspect ratio silica nanofibers," *Compos. Part B Eng.*, vol. 58, pp. 228–234, Mar. 2014, doi: 10.1016/j.compositesb.2013.10.049.

[106] Shenogin, S., L. Xue, R. Ozisik, P. Keblinski, and D. G. Cahill, "Role of thermal boundary resistance on the heat flow in carbon-nanotube composites," *J. Appl. Phys.*, vol. 95, no. 12, pp. 8136–8144, Jun. 2004, doi: 10.1063/1.1736328.

[107] Zhang, F., Q. Li, Y. Liu, S. Zhang, C. Wu, and W. Guo, "Improved thermal conductivity of polycarbonate composites filled with hybrid exfoliated graphite/multi-walled carbon nanotube fillers," *J. Therm. Anal. Calorim.*, vol. 123, no. 1, pp. 431–437, Jan. 2016, doi: 10.1007/s10973-015-4903-7.

[108] Zhang, X., L. Chen, C. Kowalski, T. Mulholland, and T. A. Osswald, "Nozzle flow behavior of aluminum/polycarbonate composites in the material extrusion printing process," *J. Appl. Polym. Sci.*, vol. 136, no. 12, p. 47252, 2019, doi: 10.1002/app.47252.

[109] Sood, M. and G. Dwivedi, "Effect of fiber treatment on flexural properties of natural fiber reinforced composites: A review," *Egypt. J. Pet.*, vol. 27, no. 4, pp. 775–783, Dec. 2018, doi: 10.1016/j.ejpe.2017.11.005.

[110] Islam, M., M. Beg, and M. Mina, "Fibre surface modifications through different treatments with the help of design expert software for natural fibre-based biocomposites," *J. Compos. Mater.*, vol. 48, no. 15, pp. 1887–1899, Jun. 2014, doi: 10.1177/0021998313491515.

[111] Hashim, M. Y., M. N. Roslan, A. M. Amin, A. M. A. Zaidi, and S. Ariffin, "Mercerization Treatment Parameter Effect on Natural Fiber Reinforced Polymer Matrix Composite: A Brief Review," *Int. J. Mater. Metall. Eng.*, vol. 6, no. 8, pp. 784–790, Aug. 2012.

[112] Brugnago, R. J., K. G. Satyanarayana, F. Wypych, and L. P. Ramos, "The effect of steam explosion on the production of sugarcane bagasse/polyester composites,"

Compos. Part Appl. Sci. Manuf., vol. 42, no. 4, pp. 364–370, Apr. 2011, doi: 10.1016/j.compositesa.2010.12.009.

[113] Jaiswal, D., G. L. Devnani, G. Rajeshkumar, M. R. Sanjay, and S. Siengchin, "Review on extraction, characterization, surface treatment and thermal degradation analysis of new cellulosic fibers as sustainable reinforcement in polymer composites," *Curr. Res. Green Sustain. Chem.*, vol. 5, p. 100271, Jan. 2022, doi: 10.1016/j.crgsc.2022.100271.

[114] Madhu, P., M. R. Sanjay, P. Senthamaraikannan, S. Pradeep, S. S. Saravanakumar, and B. Yogesha, "A review on synthesis and characterization of commercially available natural fibers: Part-I," *J. Nat. Fibers*, vol. 16, no. 8, pp. 1132–1144, Nov. 2019, doi: 10.1080/15440478.2018.1453433.

Chapter 7

The Mechanical and Thermal Behavior of Wood Flour Engineering Thermoplastic-Based Composites

Sirisart Ouajai[1,*]
and Thitinun Ungtrakul[2]
[1]Department of Industrial Chemistry, King Mongkut's University of Technology North Bangkok, Bangkok, Thailand
[2]Department of Materials and Production Engineering, The Sirindhorn International Thai-German Graduate School of Engineering (TGGS), King Mongkut's University of Technology North Bangkok, Bangkok, Thailand

Abstract

Wood flour (WF) plays a crucial role as a sustainable and cost-effective reinforcement for engineering thermoplastics. However, its influence on the thermal stability of the composite can be detrimental. This chapter delves into the key factors affecting WF composites' mechanical and thermal behavior.

The type of wood (softwood or hardwood), its chemical composition (cellulose, hemicellulose, lignin), and the size and shape of the wood flour particles all significantly impact the composite's properties. The amount of wood flour added, and any modification methods employed further influence the outcome.

Crucially, modifying wood flour can enhance its compatibility with the polymer matrix, leading to improved mechanical performance. Common modification techniques include water or solvent extraction to remove hemicellulose and the addition of compatibilizers or coupling agents.

---

* Corresponding Author's Email: sirisart.o@sci.kmutnb.ac.th

In: Mechanical and Thermal Properties of Engineering …
Editors: Senthilkumar Krishnasamy, Thitinun Ungtrakul, Sabarish Radoor et al.
ISBN: 979-8-89530-460-0

Hybrid fillers that combine wood flour with other materials can be utilized for even greater performance enhancement.
Processing and fabrication of engineering thermoplastic wood flour composites have been emphasized in exploring different mixing and compounding methods, including varying wood flour content and processing temperatures. Understanding the factors affecting the behavior of wood flour composites is not just important. It's vital. For researchers and manufacturers seeking to optimize these materials for various applications, this knowledge is the key to creating sustainable composites with the desired mechanical and thermal properties.

Keywords: wood flour, thermoplastic composite, interfacial adhesion, processing technique, thermal stability

1. Introduction

Thermoplastic resin has an advantage in reprocessing by heating and forming a product with various fabrication techniques (Biron 2013). Because the polymer chains' intermolecular connections weaken when heated, they can be repeatedly molded and reheated over multiple pressure and temperature conditions. This characteristic allows thermoplastic to be mechanically recycled into new products in large or small-scale mechanical recycling (Uzosike, Yee et al. 2023). The enhancement of thermoplastic resin property with reinforcing materials is a crucial consideration to expanding the use of commodity thermoplastic resin to the application in an engineering or specific area. In the present context, the words “engineering thermoplastics” encompass all thermoplastics that can be utilized in engineering applications. By this connotation, all commodity plastics that are appropriately filled with reinforcing or functional fillers or blended appropriately for improved performance would all be termed engineering thermoplastics (MacDermott 1997).

Due to the availability and environmentally friendly reinforcement such as natural cellulosic fibers, which are an abundant material, have been extremely attractive for sustainable production and commercialization with a reasonable cost for a thermoplastic composite. Wood fiber/flour (WF) is waste from wood mills with less contamination. Wood flour is one of the most general ingredients used to produce cellulose fiber thermoplastic composites. Regarding quality, a wide range of types, consistency for a long-term supply, and availability for large-scale or industrialized production make WF the sustainable resource for raw materials of wood polymer composite (WPC). Wood polymer composite is extensive for building, automotive, packaging, and other sectors. The application of WPC for the engineering area required careful preparation of all compositions and monitoring in every step of production to meet the mechanical property requirement of the chosen application. Thermal property is another critical property for WPC processing and service. The limitation of material like cellulose for engineering thermoplastic processing is thermal stability. Challenging this limitation has been performed with various engineering thermoplastics.

In this chapter, readers will find valuable insights on using thermoplastic composites reinforced with wood flour. The text emphasizes the influence of different wood flour characteristics like composition, surface morphology, modifier, orientation, and loading of WF on WPC properties, i.e., mechanical, thermal, and appearance. In this chapter, you will come across different types of thermoplastics. These include engineering thermoplastic composites reinforced with wood fiber (WF). This information can help researchers make informed decisions and achieve better results in WPC research.

2. Wood Flour Characteristics and Influence in Polymer Composite

Wood flour is a particle generated by manual or mechanical cutting or abrasion. Finely ground wood powder is produced by crushing, sizing, and drying a uniform product. The final product size typically ranges from 50 to 2,000 microns (Milling 2023). Milling wood flour into micron-size products is for the automotive, interior panels, and plastics industries. The quality of the wood flour significantly impacts the durability of these components. According to the materials safety data sheet, precautions are required for safe handling and storage. Wood flour is highly combustible and explosive when airborne. It is advised to store in a clean, dry, cool, well-ventilated space to prevent heat and humidity.

Wood flour is the main ingredient in wood polymer composites, so understanding fundamental knowledge about wood flour, including wood type, chemical structure, composition, morphology, surface characteristics, size, shape, and strength, is beneficial for successful composite preparation.

2.1. Appearance of Wood Flour in Polymer Composite

Wood flour is obtained from a wood mill that works on several types of wood. The color and odor of wood flour vary depending on the wood species and the period since wood dust was created. Wood flour was obtained from a local wood mill and supplier in several research studies. WPC containing wood flour is available as naturally colored granules derived from wood flour, i.e., maple, pine, oak spruce, or cedar. WPC containing maple wood flour is available as fibrous cream to a brown pellet with no odor. To achieve optimal performance of wood-plastic composites, it is essential to prioritize color consistency and durability. This is especially important for outdoor applications where the material is exposed to harsh weather conditions. Because of this, manufacturers of WPC take considerable care when choosing materials and production parameters that improve the integration of colorants, wood flour, and other constituents (BURKE 2006).

Yao Chen et al. made an exciting discovery in their research on pine wood flour. They found that by removing the extractives and dignifying them, the color and chemistry of the flour could be altered. This insight could lead to new ways of enhancing the properties of wood flour for various applications. This demonstrated that eliminating extractives had little effect on color parameters (i.e., CIE L a b), whereas delignification after extractions significantly influenced these values (Chen, Tshabalala et al. 2014). Toluene/ethanol and acetone/water extractions targeted different colored substances. Toluene/ethanol was effective for yellow-colored substances, while acetone/water mostly removed red-colored substances containing C=O groups.

2.2. Influence of Chemical Composition of Wood Flour on Thermal Stability

The surface characteristics and thermal stability of cellulose wood fiber are influenced by the noncellulosic hemicellulose that covers its surface. The composition and structure of WF's cellulose fibers determine its resistance to heat and UV radiation. Yao Chen and colleagues researched the thermal performance of pine wood flour extracted and delignified (Chen, Tshabalala et al. 2014). The process of removing lignin through delignification has been found to cause a substantial reduction in thermal stability. Increasing the onset temperature in the thermo-gravimetric analysis (TGA) thermogram was observed after removing extractives (hemicellulose), indicating enhanced thermal stability. Organo-montmorillonite (OMMT) was produced either inside or on the surface of the wood flour (WF) by treating it with sodium montmorillonite (Na-MMT) and didecyl dimethyl ammonium chloride (DDAC). The WF was modified by adding the appropriate amount of OMMT, which improved resistance to heat deformation in the composites (Liu, Peng et al. 2016).

2.3. Morphology and Size of Wood Flour

Generally, wood flour has been prepared in various sizes depending on the grinding process, source of raw materials, and further fabrication of WPC techniques. Mechanical grinding of wood produces particles with various shapes, small sizes, and high surface area. Cellulose crystallinity may be low or high depending on the raw material properties, grinding conditions, energy, and mechanism applied. Wood flour's type and size significantly affect WPC's physical and mechanical properties and limit the fabrication technique, i.e., 3D printing (Petchwattana, Channuan, et al., 2019).

Wood flour size reduction with different mills uses different fracture mechanisms. The dense natural deposition of fibrils within a single WF particle bound together with lignin and/or other noncellulosic components is pulverized to partial or complete separation of fibrils (Haque, Goda et al. 2019).

The composite's properties are influenced by the surface morphology and degree of defibrillation of wood flour, as demonstrated by Md Minhaz-Ul Haque et al. (Haque, Goda, et al. 2019). When wood flour is fibrillated and pulverized with water, it increases the roughness of the particle surfaces. Introducing wet pulverized wood flour can enhance the mechanical properties of PP/WF, including its tensile strength, impact strength, and fatigue life. Composites have a longer fatigue life, indicating that fibrillated wood flour particles by wet pulverization have better crack resistance.

The findings of P. Karinkanta et al. are noteworthy as they have collected data from multiple researchers who have studied the aspect ratio of wood powder concerning median particle size. This information is valuable in understanding the properties of wood powder and can aid in the improved process. It was found that the jet mill produced the highest aspect ratio compared to other mills. A higher aspect ratio of wood flour is typically associated with a higher median particle size. This result correlated well with a study investigating the impact of wood flour distribution on WPCs' tensile and flexural behavior (Karinkanta, Ämmälä et al. 2018). In general, composites' mechanical properties are improved by adding large particles, which have a higher aspect ratio (Chaudemanche, Perrot et al. 2018). Due to a higher aspect ratio, the improved stress transfer to the wood flour is the main contributor to the higher modulus of elasticity values in WPC. The size distribution of wood flour particles affected the physical and mechanical properties of a wood plastic composite processed via extrusion. The orientation of wood particles varies depending on the size of the scattering of the wood flour used. The extrusion direction (MD) aligns large particles, improving the flexural and tensile capabilities of the coarse wood flour PE composite. The extrusion direction (MD) exhibits significantly higher elastic modulus and tensile strength due to the orientation of wood fibers during extrusion, compared to the transversal direction. Local deformation processes may occur in the PP/elastomer/wood composite. The type and amount of processes are incredibly dependent on particle size and interfacial adhesion. Large wood particles show a large number of debonding events. Small wood particles were also deboned, but not in significant quantities, because debonding stress increases as particle size decreases (Pukanszky and VÖRÖS 1993).

2.4. Loading of Wood Flour

Adding wood flour to the composite facilitates load transfer from the matrix to the WF fiber. Generally, the composition of wood flour in composites ranged from 0.1 to 0.3 volume fractions. An increase in WF volume fraction improved Young's modulus of PP composite. (Pérez, Famá et al. 2012, Ferdinánd, Jerabek et al. 2023) The research findings indicate a clear inverse correlation between the WF volume fraction and the ultimate tensile strength; an increase in the former results in a decrease in the latter. The observed reduction in tensile strength and ductility of the untreated samples is due to embrittlement initiated by improper fiber wetting. This leads to an inconsistent fiber matrix, which could severely affect the material's structural stability. Primary energy-absorbing mechanisms found in the unmodified composites were debonding and fiber pull-out.

As the concentration of wood flour increased, the crystallinity of the neat PHBV matrix decreased. (Srubar, Pilla et al. 2012) Due to the addition of wood flour to the PHBV matrix, a smaller perfect arrangement of polymer crystal structures occurs. These defects and irregularities may lead to fractures in crystalline polymers. Unlike continuous fibers, wood fibers lack sufficient lengths to activate toughening processes that prevent brittle fracture.

2.5. Modification of Wood Flour for Interfacial Enhancement and Thermal Stability Improvement

The key to achieving the desired mechanical strength in WPC (wood plastic composites) is improving the interaction between wood fibers and a hydrophobic polymer matrix. To study the influence of the modifier type and composition on the property of WPC, wood fiber pre-treatment (such as hot water, solvent extraction, and alkalization) and chemical addition as a compatibilizer or modifier have been frequently used. This section will provide an example of research that aims to enhance the compatibility between wood fiber surfaces and matrixes.

2.5.1. Influence of Noncellulosic Component Removal

The study of Gardner Oporto aimed to evaluate the compatibility of surfaces in wood-plastic composites (WPCs) made with polypropylene and styrene maleic anhydride. The assessment used various hardwood and softwood flour types, including hot water-extracted and un-extracted and extracted hardwood flour. The method involved measuring the work of adhesion and cohesion for different wood-based reinforcing materials and thermoplastic resins through IGC retention time measurements after determining their surface energy (Oporto, Gardner et al. 2011). The chemical composition of hardwood and softwood contains hemicellulose, which affects the material's surface energy. Glucomannans increase the surface energy, while xylans decrease it. When hardwood is processed to extract hemicelluloses, it increases surface energy. This is due to the negative influence of xylans on surface energy. A recent study discovered that the dispersion component of the surface energy for materials such as polypropylene, styrene maleic anhydride (SMA), and resin-based materials was similar to that of wood-based materials. However, significant differences were observed in the acid-base component. This difference was expected, given the chemical composition of the materials. The study found that SMA-containing resin had a higher surface energy, while the nonpolar backbone structure of polypropylene had low surface energy. The composite's mechanical properties increased as the adhesion work to cohesion ratio increased. Similarly, the thermoplastic resin's higher surface energy acid-base component resulted in improved properties.

The initial processes for enhancing purity and promoting the interfacial interaction of wood flour with the polymer matrix are water steaming, solvent extraction, and alkalization. WF is steamed in a boiling pot to prepare braised wood flour (BWF). The braising wood flour is highly effective in enhancing the mechanical properties and melt viscosity of composites made of recycled polyethylene and wood flour. It works by expanding the gaps between cells, which leads to better infiltration of recycled polyethylene into the gaps, resulting in improved properties and performance of the composites (Hong, Guo et al., 2019).

Yao Chen and their colleagues researched to examine the influence of extractives and lignin on the thermal stability of wood flour. They utilized the thermogravimetry technique to investigate the thermal stability of mixed pine wood flour extracted and delignified. The outcomes of the study indicated that lignin had a more significant effect on the thermal stability of the sample than extractives. They also found that increasing the polarity of solvents improved thermal stability by raising the onset temperature T_0 by removing extractives. The study revealed that Wood-Plastic Composites produced with high-density polyethylene and extracted wood flour exhibited enhanced mechanical properties and lower water absorption rates than those made with untreated wood flour. These findings could prove valuable in developing more durable and water-resistant composite materials (Chen, Tshabalala et al. 2014).

A new approach was explored to improve compatibility between wood flour and the polymeric matrix of wood-plastic composites, led by Huiping Lin and a group of researchers. They aimed to make WF more hydrophobic without using organic modifiers or compatibilizers. The team treated WF with p-toluenesulfonic acid (p-TsOH) to selectively remove hemicellulose, significantly increasing the material's hydrophobicity. The composite containing 50% MWF proved to have the best mechanical properties among the samples tested. It demonstrated an impressive 33.5% increase in tensile strength and a remarkable 38.7% increase in flexural strength compared to the WF/HDPE composite. These results highlight the potential for further optimizing the composite's properties by adjusting the concentration of MWF in the mixture.

After treating the beech fibers with sodium hydroxide (NaOH) solution, their adhesion to the PA 11 matrix improved. Removing hemicellulose from the fibers resulted in increased thermal stability, as Thermogravimetric

Analysis (TGA) indicated. The extrapolated TGA onset temperature increased from 290°C to 330°C (Zierdt, Theumer et al. 2015).

2.5.2. Influence of Chemical Modifiers

Different additives improved the bonding between the wood flour and the polymer matrix. The fibers were detached from the matrix without these additives due to poor bonding. Maleated PP (MAPP) was applied to improve the adhesion. Using MAPP led to better tensile strength and flexibility without significantly affecting the fracture toughness. This is because MAPP improved the dispersion of fibers within the matrix and the bonding between the wood flour and polymer (Pérez, Famá, et al. 2012).

The use of polyethylene ternary graft copolymers (GPE) as compatibilizers in rPE/WF composites has been studied. A GPE with a graft degree of 10.5% was produced by grafting MAH, methyl methacrylate, and butyl acrylate in a solid state. The rPE/BWF composites, compatible with GPE, exhibited superior interfacial interactions (Hong, Guo, et al., 2019).

Adding 1% maleic anhydride (MA) dramatically enhances the mechanical properties of PLA/wood flour composites. The modified composite exhibits a significant improvement in both tensile and flexural strength properties. According to recent studies, the modified composite showed a remarkable 144% increase in tensile strength, while the flexural strength increased by 44%. This improvement was due to incorporating anhydride groups, which improved the compatibility between PLA and wood flour. The anhydride groups facilitated stronger interfacial adhesion by reacting with hydroxyl groups, ultimately enhancing the material's overall performance. Thus, the modified composite has the potential to be used in various structural applications that require high mechanical strength (Lv, Gu et al. 2016).

The study conducted by Cecilia Zárate-Pérez et al. was aimed at improving the mechanical and thermal properties of PP/WF composites by introducing coupling agents such as grafted copolymers based on Poly(styrene–glycidyl methacrylate)-GMA- or Poly(styrene-co-methyl methacrylate-co-glycidyl methacrylate)-MMA-. The results were promising, as the use of coupling agents had a significant positive impact on the mechanical behavior of the composites. In particular, the elastic modulus of the composites increased by approximately 82%, which is a noteworthy improvement compared to Young's modulus of PP (Zárate-Pérez, Ramírez-Aguilar, et al. 2023).

As a coupling agent, polymeric diphenylmethane diisocyanate (pMDI) was applied for WPC from PBS and WF. The PBS composite's tensile strength was enhanced as 2phr pMDI was incorporated (Park, Youe, et al., 2019). It has been discovered that pMDI can significantly improve the tensile properties of polyhydroxy butyrate (PHB)/wood flour (WF) composites. Adding pMDI resulted in strong adhesion between the PHB and the fiber. In composites modified with 4% pMDI, the fiber surface showed a clean fracture instead of fiber pull-out. This indicates that the fracture likely started from the matrix and propagated directly through the fiber. The incorporation of MDI proved a valuable addition to WF/TPU, leading to a significant improvement of 22.33% in its tensile strength.

A study by Hongjie Bi and colleagues revealed that the WF/TPU composite, modified with EPDM-g-MAH, exhibited the highest elongation at break. The elongation increased by approximately 100.14% compared to the unmodified composite. The study also found that the EPDM-g-MAH and WF had good compatibility. Their interaction was possible because of the reaction between maleic anhydride and OH groups. EPDM acted as a cross-linker in TPU's main chain (Bi, Ren et al. 2018).

It was found that adding MAPO to PP/WF composites did not enhance their fatigue performance. This indicates that PP's molecular degradation, as indicated by rheological measurement, has a more significant impact on the composites' fatigue performance than the compatibilizing effect of MAPO (Haque, Goda et al. 2019).

The introduction of MAPP into the composite material significantly enhanced the material's ability to resist pulling forces and increased its flexibility while leaving its ability to withstand cracks and fractures unchanged. In the unmodified composites, energy-absorbing mechanisms primarily involved debonding and fiber pull-out, whereas in the composites with MAPP, fiber breakage was the dominant mechanism due to enhanced interfacial adhesion (Pérez, Famá et al. 2012).

2.6. Hybrid Filler with Wood Flour

Researchers have introduced additional fillers and reinforcements further to enhance the thermal and mechanical properties of WPCs. This has opened up possibilities for WPCs' use in various industries and applications requiring specific material properties.

Wood-plastic composites (WPCs) are more prone to catching fire than wood because their thermoplastic plastics have a high heating value. Developing WPCs with good flame resistance is crucial to reduce the risk of fire. To achieve this, a flame-retardant polypropylene/wood flour (PP/WF) composite was created by melt blending aluminium hypophosphite (AP) and melamine cyanurate (MCA). It was observed that the inclusion of flame retardant reduced the tensile strength of PP/WF composites, but the flexural strength of the composite increased due to the compatibility between PP and MCA and the lubricity of AHP and MCA (Zhao, Guo et al., 2019).

To enhance the reinforcement efficacy of wood/HDPE composites, a study was conducted to investigate the effects of particle size and dispersion on the composites. This was achieved by incorporating silica particles of micro and nano sizes into high-density polyethylene (HDPE) using two different methods: direct compounding (dry dispersion) and mixing in a solution (wet dispersion). The results of this study can help determine the most effective approach for improving the reinforcing properties of wood/HDPE composites. Adding silicate particles enhanced wet dispersion patterns, resulting in a maximum improvement of 16%, 18%, and 30% in flexural, tensile, and impact strength, respectively. Using inorganic particles like silicon enhances WPCs' mechanical properties (Zhou, Hao et al., 2018).

Some additives can be added to decayed wood plastic composites (DWPC) to create a more durable and environmentally friendly material. These additives include nano calcium carbonate (NCC), activated carbon (AC), and chitosan (CS). By adding these additives to DWPC, a longer-lasting and eco-friendly material can be created. Soundwood plastic composites (SWPC) were also prepared without any added substances. Incorporating CS in DWPCs has been found to impact various aspects of polymer performance positively. Compared to SWPCs and DWPCs without additives, DWPCs with CS demonstrate improved mechanical properties, such as strength and durability, enhanced chemical structure, thermal stability, degree of crystallinity, and flame retardancy. This suggests that using CS as an additive in DWPCs can offer significant overall performance and safety benefits. Therefore, the addition of CS can be considered a valuable approach to enhancing the properties of DWPCs (Ge, Zuo et al. 2021).

3. Processing of Thermoplastic Wood Flour Composite

Since the processing temperature of engineering thermoplastic is around 200-300 degrees Celsius, this is too high for composite preparation containing general cellulose fiber. Improving thermal stability is the prior requirement for wood flour for composite application. In addition, the mechanical property is high, so the inclusion of cellulose fiber must be considered in a strong structure and interaction with engineering polymer.

3.1. Mixing and Compounding

The mixing or compounding temperature is a crucial processing parameter affecting Wood Flour (WF) degradation and the matrix that undergoes elevated temperature. An internal mixer was used to create composites with different proportions of wood flour. The wood flour was added to the mixer at varying weights, with 10%, 20%, and 30% used. The mixer was set to operate at a temperature of 180°C and a speed of 45 rpm, and the materials were mixed for 10 minutes. This process allowed for the creation of composites with different levels of wood flour, each possessing unique properties and characteristics (Pérez, Famá, et al. 2012).

The extrusion process was the main operation for wood flour polymer composite preparation. The maximum set temperature of 210°C was found for PP and 165 °C for HDPE, respectively. A one-step process and 2-step process were conducted and aimed to prepare a good dispersion of WF. In addition, a dry and wet process was employed to compare the effects of the process on the dispersion of the WF inter matrix (Zhou, Hao et al., 2018).

The thermal stability of wood plastic composites is an essential parameter that determines the success of the processing step. High temperatures and shear rates are applied during the extrusion process, which can cause a significant degradation process. This degradation can lead to a decline in the physical and mechanical properties of the composites, such as strength, stiffness, and durability. Therefore, it is crucial to ensure that the thermal stability of the composites is optimized to maintain their integrity and quality during the processing step. It is essential to adjust the extrusion parameters based on thermal stability to create high-quality composites using wood flour. To make a PVB/WF composite, a single screw extruder was used, which had a Maillefer screw with a mixer block-type pineapple located in the metering zone. The composite was mixed at varying weight percentages of 25, 40, and 55 wt%. This mixing process involves the extruder forcing the PVB/WF composite through the mixer block, which contains the pineapple-shaped elements that facilitate the mixing and dispersion of the wood fibers within the polymer matrix. The extrusion was operated in a starved feed condition in its feeding zone. There is a good dispersion of wood flour and adhesion between

PVB and wood in composites with higher wood fraction, even though wood aggregates are present (Ambrósio, Sônego et al. 2019).

The PLA/PBS/WF composites were produced using a 20mm co-rotating twin-screw extruder with a 32 L/D ratio (LW-20-32, Labtech Engineering Company). The extruder was operated with a temperature profile ranging from 165°C to 180°C, with the highest temperature at the die. The composite morphology showed distinct boundaries and gaps between the WF and matrix, indicating that the hydrophilicity difference between WF and polyesters did not mix well. As the amount of WF increased, the WF aggregation became more evident, and the interface between the WF and the PLA/PBS matrix became clearer (Chuayjuljit, Wongwaiwattanakul et al. 2017).

3.2. Fabrication of Composite

Various fabrication techniques were employed for WPC reinforced with wood flour. A wood flour thermoplastic compound prepared from an extrusion process was further fabricated to test the specimen and product. The favorite and novel techniques currently used in WPC fabrication are compression molding, injection molding, and additive manufacturing. However, this section will focus on only additive manufacturing (AM) technology for WPC, the latest discovery in this area.

Additive manufacturing, widely known as 3D printing, is a groundbreaking technology transforming the manufacturing industry. This process simplifies the precision of manufacturing intricate 3D objects. Various kinds of 3D printing processes exist. Nevertheless, fused deposition modeling (FDM) and Laser Sintering (LS) are explored in WPC preparation.

Fused Deposition Modelling (FDM) is a widely used and dependable form of additive manufacturing technology. It is renowned for its affordability, ease of use, minimal wastage, and material availability. WPC has received attention in 3D printing with an effort to develop more ecological manufacturing methods. The extrusion process fabricates wood flour composite filaments for fused deposition modeling (FDM). Several critical factors, such as the amount of wood content, particle/fiber geometry, and particle size, determine the printability of WPC filament. To prevent clogging at the printing nozzles, the particle size of the fiber particles should be small enough. According to Petchwattana et al., composite materials with a particle diameter of 70 μm are the only ones that can be printed successfully (Petchwattana, Channuan, et al., 2019).

The inclusion of WF in the PLA matrix disrupted the crystallization process. The presence of WF constrained the PLA matrix's ability to crystallize, leading to a decline in the glass transition temperature and cold crystallization temperature from 67 to 60°C and 101 to 97°C, respectively (Tao, Wang et al. 2017).

Bi, Ren, et al. investigated the use of TPU/WF in preparing 3D printing products. The objective was to examine the impact of various WF contents and modifiers on composite material characteristics. Upon completing a 3D printing experiment, it was discovered that the mobile phone shell design had a smooth surface, a texture resembling natural wood, and exceptional flexibility. This sets the product apart from conventional and fragile 3D printing materials (Bi, Ren et al. 2018).

Laser sintering (LS) is a technology that offers a cost-effective and eco-friendly solution. A eucalyptus-polyamide 12 (EPA12) composite has been developed and tested for its strength and modulus using various laser powers, focusing on sustainability. The results indicated a significant increase in tensile strength and Young's modulus with increased laser power. There is a direct correlation between density and tensile strength. Tensile strength and density are directly correlated with one another. When a high laser power is applied to eucalyptus fibers, it causes the Polyamide 12 (PA12) to wet the fibers more extensively, thereby improving the density, bonding strength, tensile strength, and Young's modulus of the part. This happens because PA12 particles fuse more effectively, and there is less porosity, which results in a more compact structure (Jiang, Guo et al. 2014).

4. Engineering Thermoplastic Wood Flour Composite

Limited engineering thermoplastic wood flour composites were found to have a high processing temperature. Nevertheless, some attempts were made to treat wood flour with special treatment to withstand high-temperature processing in a few engineering thermoplastics, namely polyamide and polycarbonate.

4.1. Polyamide WF Composite

Composites with wood flour filler had better properties than rice husk-filled composites at 20-30% filler content. The presence of lignocellulosic filler reduced impact strength values. The melting temperature and crystallinity of the W-PA6G-based composites were reduced with lignocellulosic filler. The decrease in melting temperature and crystallinity was more distinct in the RH-filled composites. The ones that were filled with RH showed a more noticeable decrease (Şeker Hirçin, Yörür et al. 2021).

A treatment with boric acid can improve the thermal stability of lignocellulose to prepare bio-composites based on engineering plastic. Boric acid protects lignocellulose from thermal degradation, increasing the lightness of the resulting composites. The addition of boric acid negatively impacted the mechanical strength of the resulting composites. However, boric acid seemed to have a detrimental result on the composites' mechanical strength (Zhang, Koubaa, et al., 2020).

As a result of the chemical treatment of fibers in the bio-WPC, the mechanical and thermo-mechanical analyses demonstrated an increase in the composites' elastic and storage modulus. As a result of the chemical treatment of the fibers, mechanical and thermo-mechanical studies of the processed bio-WPC revealed an increase in the composites' elastic and storage moduli. The enhanced performance of the composite material can be attributed to the formation of a tremendous amount of hydrogen bonds among the modified beech fibers and the PA 11 matrix. This constructive bonding has resulted in a material that is stronger, more durable, and better equipped to handle challenging conditions (Zierdt, Theumer et al. 2015).

Fabien Sliwa and his team studied novel wood thermoplastic elastomer composites that use a PEBAX copolymer as matrices instead of the traditional WPC matrices. These copolymers are polyether-b-amide thermoplastic elastomers with the necessary elongation at break and a melting point below 200°C. This is crucial to avoid the degradation of wood fiber during compounding. These polymers are derived from renewable resources and possess a hydrophilic nature that enables them to interact with wood fibers (Sliwa, El Bounia et al. 2012).

4.2. Polycarbonate WF Composite

The utilization of bio-based composites in commercial settings is a subject of concern due to the potential fire risk. However, it is essential to acknowledge that bio-based composites are a valuable resource in reducing carbon emissions and promoting sustainability. With careful consideration and implementation of safety measures, businesses can continue incorporating bio-based composites while ensuring safety. Zhang J et al. enhanced the thermal stability and flame retardancy of wood flour by treating it with boric acid before combining it with PC and high-density polyethylene (HDPE). All compositions were mixed at a maximum temperature of 220 °C. Adding PC lowered the heat release rate of the composites and slowed the composites' heat release rate during combustion. The resulting composites' tensile strength and modulus of elasticity increased by 27.7% and 91.1%, respectively (Zhang, Koubaa, et al. 2021).

Zhang J et al. investigated the impact of incorporating borated wood flour on the fire retardancy and water absorption properties of wood flour/PC composites to enhance their applicability. Adding wood flour increased the fire retardancy of the composites and resulted in a 102.3% increase in their char residue compared to pure PC. The composites' char residue rose by 102.3 percent compared to pure PC, and their fire retardancy improved as the amount of wood flour increased. Hydroxyl groups in the wood flour increased water absorption when incorporated into composites. Despite this, the composites showed improved anti-creep properties, which is highly desirable in structural elements subjected to long-term loads. Adding wood flour enhanced the mechanical behavior of the composites by reinforcing their matrix, ultimately leading to their improved anti-creep performance. Wood-PC composites' outstanding fire retardancy and anti-creep characteristics expand their utility in the construction industry.

Conclusion

The success of wood flour polymer composites (WPCs) depends on multiple critical parameters, including wood flour orientation, dispersion quality in the matrix, particle size and aspect ratio, loading percentage, interfacial adhesion, thermal stability, and processing techniques. Controlling these factors is essential for achieving the desired properties in the final composite material.

Wood flour's inherent characteristics, including its source (softwood or hardwood), chemical composition (cellulose, hemicellulose, lignin), and particle morphology, significantly influence the composite's mechanical and thermal performance. Pre-treatment methods, such as hot water extraction, solvent extraction, and alkalization, can remove non-cellulosic components, enhancing thermal stability and compatibility with polymer matrices.

Surface modification of wood flour through chemical treatments and compatibilizers has proven effective in improving interfacial adhesion with hydrophobic polymer matrices. Coupling agents like maleated polypropylene

(MAPP), polyethylene ternary graft copolymers (GPE), and polymeric diphenylmethane diisocyanate (pMDI) create stronger bonds between wood flour and various polymer matrices, significantly enhancing mechanical properties.

Introducing hybrid fillers that combine wood flour with materials like aluminium hypophosphite, silica particles, nano calcium carbonate, and chitosan offers additional pathways to improve specific properties such as flame retardancy, mechanical strength, and thermal stability. These innovations expand the potential applications of WPCs in various industries.

Processing parameters, particularly temperature control during mixing and fabrication, are crucial to prevent the thermal degradation of wood components while ensuring proper dispersion. Advanced fabrication techniques like additive manufacturing (3D printing) create new opportunities for WPC applications, though they require careful consideration of particle size to prevent nozzle clogging.

Developing WPCs using engineering thermoplastics like polyamide and polycarbonate demonstrates the potential to extend these composites into more demanding applications. However, special treatments are required to withstand wood flour's high processing temperatures.

For optimal WPC performance, manufacturers must carefully select wood flour quality, optimize processing parameters, and choose appropriate pre-treatments and surface modifiers that match the specific polymer matrix. These considerations can help develop sustainable, high-performance composite materials with tailored properties for various applications.

References

Ambrósio, J. D., M. Sônego, L. H. Staffa, M. A. Chinelatto and L. C. Costa (2019). "Characterization of flexible Poly(vinyl butyral)/wood flour composites." *Composites Part B: Engineering* 175: 107118.

Bi, H., Z. Ren, R. Guo, M. Xu and Y. Song (2018). "Fabrication of flexible wood flour/thermoplastic polyurethane elastomer composites using fused deposition molding." *Industrial Crops and Products* 122: pp. 76–84.

Biron, M. (2013). 5 - *Thermoplastic Processing. Thermoplastics and Thermoplastic Composites* (Second Edition). M. Biron, William Andrew Publishing: 715–768.

Burke, J. (2006). *Fixing the Mixing in Wood-Plastic Profiles: Color consistency and durability are critical to wood-plastic composites, especially in outdoor applications.*

Chaudemanche, S., A. Perrot, S. Pimbert, T. Lecompte and F. Faure (2018). "Properties of an industrial extruded HDPE-WPC: The effect of the size distribution of wood flour particles." *Construction and Building Materials* 162: 543-552.

Chen, Y., M. Tshabalala, J. Gao, N. Stark and Y. Fan (2014). "Color and Surface Chemistry Changes of Pine Wood Flour after Extraction and Delignification." *Bioresources* 9: 2937.

Chen, Y., M. A. Tshabalala, J. Gao, N. M. Stark, Y. Fan & R. E. Ibach (2014). "Thermal behavior of extracted and delignified pine wood flour." *Thermochimica Acta* 591: 40-44.

Chuayjuljit, S., C. Wongwaiwattanakul, P. Chaiwutthinan & P. Prasassarakich (2017). "Biodegradable poly(lactic acid)/poly(butylene succinate)/wood flour composites: Physical and morphological properties." P*olymer Composites* 38(12): 2841–2851.

Ferdinánd, M., M. Jerabek, R. Várdai, T. Lummerstorfer, C. Pretschuh, M. Gahleitner, G. Faludi, J. Móczó and B. Pukánszky (2023). "Impact modification of wood flour reinforced PP composites: Problems, analysis, solution." *Composites Part A: Applied Science and Manufacturing* 167: 107445.

Ge, S., S. Zuo, M. Zhang, Y. Luo, R. Yang, Y. Wu, Y. Zhang, J. Li & C. Xia (2021). "Utilization of decayed wood for polyvinyl chloride/wood flour composites." *Journal of Materials Research and Technology* 12: pp. 862–869.

Haque, M. M.-U., K. Goda, H. Ito, S. Ogoe, M. Okamot, T. Ema, K. Kagawa and H. Nogami (2019). "Melt-viscosity and mechanical behaviour of polypropylene (PP)/wood flour composites: Effect of pulverization of wood flour with and without water." *Advanced Industrial and Engineering Polymer Research* 2(1): 42-50.

Haque, M. M.-U., K. Goda, S. Ogoe and Y. Sunaga (2019). "Fatigue analysis and fatigue reliability of polypropylene/wood flour composites." *Advanced Industrial and Engineering Polymer Research* 2(3): 136-142.

Hong, H., Q. Guo, H. Zhang and H. He (2019). "Effect of interfacial modifiers and wood flour treatment on the rheological properties of recycled polyethylene/wood flour composites." *Progress in Rubber, Plastics and Recycling Technology* 36(1): 31–46.

Jiang, K., Y. L. Guo and D. L. Bourell (2014). *Processing and Characterizations of Eucalyptus-PA12 Composite by Laser Sintering.*

Karinkanta, P., A. Ämmälä, M. Illikainen and J. Niinimäki (2018). "Fine grinding of wood – Overview from wood breakage to applications." *Biomass and Bioenergy* 113: 31-44.

Liu, R., Y. Peng and J. Cao (2016). "Thermal stability of organo-montmorillonite-modified wood flour/poly(lactic acid) composites." *Polymer Composites* 37(7): 1971-1977.

Lv, S., J. Gu, H. Tan & Y. Zhang (2016). "Modification of wood flour/PLA composites by reactive extrusion with maleic anhydride." *Journal of Applied Polymer Science* 133(15).

MacDermott, C. P. (1997). *Selecting Thermoplastics for Engineering Applications. Selecting Thermoplastics for Engineering Applications.* C. P. MacDermott and A. V. Shenoy, CRC Press.

Milling, M. (2023). "What is wood flour." Retrieved 27 December 2023, from http://www.micromilling.com.au/woodflour.

Oporto, G. S., D. J. Gardner, A. Kiziltas and D. J. Neivandt (2011). "Understanding the Affinity between Components of Wood–Plastic Composites from a Surface Energy Perspective." *Journal of Adhesion Science and Technology* 25(15): 1785-1801.

Park, C.-W., W.-J. Youe, S.-Y. Han, J.-S. Park, E.-A. Lee, J.-Y. Park, G.-J. Kwon, S.-J. Kim and S.-H. Lee (2019). "Influence of Lignin and Polymeric Diphenylmethane Diisocyante Addition on the Properties of Poly(butylene succinate)/Wood Flour Composite." *Polymers* 11(7): 1161.

Pérez, E., L. Famá, S. G. Pardo, M. J. Abad and C. Bernal (2012). "Tensile and fracture behaviour of PP/wood flour composites." *Composites Part B: Engineering* 43(7): 2795-2800.

Petchwattana, N., W. Channuan, P. Naknaen & B. Narupai (2019). "3D Printing Filaments Prepared from Modified Poly(Lactic Acid)/Teak Wood Flour Composites: An Investigation on the Particle Size Effects and Silane Coupling Agent Compatibilisation." *Journal of Physical Science* 30: pp. 169–188.

Pukanszky, B. and G. VÖRÖS (1993). "Mechanism of interfacial interactions in particulate filled composites." *Composite Interfaces* 1(5): 411-427.

Şeker Hirçin, B., H. Yörür and F. Mengeloğlu (2021). "Effects of filler type and content on the mechanical, morphological, and thermal properties of waste casting polyamide 6 (W-PA6G)-based wood plastic composites." *BioResources* 16(1): 655-668.

Sliwa, F., N.-e. El Bounia, F. Charrier, G. Marin and F. Malet (2012). "Mechanical and interfacial properties of wood and bio-based thermoplastic composite." *Composites Science and Technology* 72(14): 1733-1740.

Srubar, W. V., S. Pilla, Z. C. Wright, C. A. Ryan, J. P. Greene, C. W. Frank & S. L. Billington (2012). "Mechanisms and impact of fiber–matrix compatibilization techniques on the material characterization of PHBV/oak wood flour engineered biobased composites." *Composites Science and Technology* 72(6): 708–715.

Tao, Y., H. Wang, Z. Li, P. Li and S. Q. Shi (2017). "Development and Application of Wood Flour-Filled Polylactic Acid Composite Filament for 3D Printing." *Materials* 10(4): 339.

Uzosike, C. C., L. H. Yee and R. V. Padilla (2023). "Small-Scale Mechanical Recycling of Solid Thermoplastic Wastes: A Review of PET, PEs, and PP." *Energies* 16(3): 1406.

Zárate-Pérez, C., R. Ramírez-Aguilar, E. A. Franco-Urquiza and C. Sánchez-Alvarado (2023). "The Role of Coupling Agents in the Mechanical and Thermal Properties of Polypropylene/Wood Flour Composites." *Macromol* 3(1): 65-78.

Zhang, J., A. Koubaa, D. Xing, W. Liu, Q. Wang, X. Wang and H. Wang (2020). "Improving lignocellulose thermal stability by chemical modification with boric acid for incorporating into polyamide." *Materials & Design* 191: 108589.

Zhang, J., A. Koubaa, D. Xing, H. Wang, F. Wang, X.-M. Wang and Q. Wang (2021). "Flammability, thermal stability, and mechanical properties of wood flour/polycarbonate/polyethylene bio-based composites." *Industrial Crops and Products* 169: 113638.

Zhao, P., C. Guo and L. Li (2019). "Flame retardancy and thermal degradation properties of polypropylene/wood flour composite modified with aluminum hypophosphite/melamine cyanurate." *Journal of Thermal Analysis and Calorimetry* 135(6): 3085–3093.

Zhou, H., X. Hao, H. Wang, X. Wang, T. Liu, Y. Xie & Q. Wang (2018). "The reinforcement efficacy of nano- and microscale silica for extruded wood flour/HDPE composites: the effects of dispersion patterns and interfacial modification." *Journal of Materials Science* 53.

Zierdt, P., T. Theumer, G. Kulkarni, V. Däumlich, J. Klehm, U. Hirsch and A. Weber (2015). "Sustainable wood-plastic composites from bio-based polyamide 11 and chemically modified beech fibers." *Sustainable Materials and Technologies* 6: 6-14.

Chapter 8

Revolutionizing Healthcare and Dentistry: The Transformative Impact of 3D Printing on Customized Solutions

K. Senthilkumar[1,*]
D. Aravind[2]
M. Satthiyaraju[3]
J. Jerold John Britto[4]
S. Balakrishnan[5]
T. Senthil Muthu Kumar[6]
and M. Chandrasekar[7]
[1]Department of Mechanical Engineering, PSG Institute of Technology and Applied Research, Tamil Nadu, India
[2]University Science Instrumentation Centre, Madurai Kamaraj University, Palkalai Nagar, Tamil Nadu, India
[3]Department of Mechanical Engineering, SRM Institute of Science and Technology, Vadapalani Campus, Tamilnadu, India
[4]Department of Mechanical Engineering, Ramco Institute of Technology, Tamilnadu, India
[5]Department of Mechanical Engineering, PPG Institute of Technology, Tamil Nadu, India
[6]Department of Mechanical Engineering, Kalasalingam Academy of Research and Education, Tamil Nadu, India
[7]School of Aeronautical Sciences, Hindustan Institute of Technology & Science, Padur, Kelambakkam, Tamil Nadu, India

---

[*] Corresponding Author's Email: kmsenthilkumar@gmail.com

In: Mechanical and Thermal Properties of Engineering …
Editors: Senthilkumar Krishnasamy, Thitinun Ungtrakul, Sabarish Radoor et al.
ISBN: 979-8-89530-460-0

Abstract

This review article explores the use of 3D printing in the biomedical field. Although various 3D printing techniques exist, fused deposition modelling (FDM) is an exceptionally versatile choice for crafting biomedical prototypes and structures. The utilization of 3D printing offers many advantages in the biomedical field, particularly in dental implants. Notably, they can provide patient-specific designs and unique implant structures. To utilize the full potential of the FDM technique in any field, including biomedical industries, it is imperative to understand its mechanical properties. Factors such as the gap sizes between adjacent paths and filament orientations influence mechanical performance. For instance, smaller gaps can enhance the material density, whereas meticulous filament orientation optimization enhances structural integrity. This review article provides an overview of the role of 3D printing in biomedical applications, including mechanical properties and dental prototypes. A promising future for this technology is to improve patient care and outcomes.

Keywords: 3D priting, biomedical, mechanical properties, dental

1. Introduction

3D printing is the widely applied additive manufacturing (AM) technique for numerous fields such as materials engineering, mechanical engineering, computer science, electrical, electronics, biomedical engineering etc. The evolution of 3D printing started as early as the 1980s. 3D printing was marketed 3D-Systems for the first time in 1987 and is currently applied not only for fabricating prototypes but also for large-area manufacturing. The advantages of 3D printing are less material loss, ease of fabricating complex shapes and cost effective [1].

3D printing has matured as a driving force for the world's development in the technology for manufacturing, and hence they are implied widely in the sectors of space, automotive parts, bio-medicals prototypes, home accessories, stationeries, ornamental making, creative items etc. They are simply accessible, effective, efficient as they are not complex, cost-effective and quick. 3D printing involves the fabrication of products directly from their raw material using a 3D digital model, as the raw material gets added in a layer-by-layer sequence [2]. The FDM technique is the widely used manufacturing technique of all 3D printing techniques because of its non-laser application. It is cheaper in terms of usage and maintenance. Besides, the equipment set up for FDM is smaller, portable, and easy to use with a common operating interface. Also, the fabrication that requires multi-phase to get multi-functional compound materials can be achieved through multiple nozzles and multichannel. A magnificent achievement in the field 3D printers was made when the FDM-named additive manufacturing facility (AMF) was sent to space for the first time by the US National Aeronautics and Space Agency (NASA) in 2014 to print the spanner used for maintenance and repair in the orbit's space station [3, 4].

A wide range of materials was developed for 3D printing, including synthetic polymers, natural polymers, composites, metals, ceramic substances and wax material [5–8]. Polymer filaments with a low melting point are the consumable 3D printing materials, which mostly includes polylactic acid (PLA), acrylonitrile butadiene styrene (ABS), polycarbonate (PC), and polyamide (PA). Though materials like acrylonitrile butadiene styrene (ABS), thermoplastic polyurethane (TPU), nylon, and polyetheretherketone (PEEK) are the most preferred 3D printing materials, they are potentially hazardous too. This could be because of harmful volatile organic compounds and ultrafine aerosol that are found released during 3D printing processes. For example, the 3D-printed objects fabricated using these polymers are found to be toxic to zebrafish embryos, which are model organisms widely used for research in biological labs [9]. Therefore, it is essential to manufacture hazardous-free 3D printing materials that are less toxic.

Natural polymers have become important and replaced environmentally risky synthetic polymers as the most sought-after 3D printing material [10]. For example, PLA, one of the promising and widely used FDM 3D printing material technology, has gained the interest of researchers. Polylactic acid (PLA) is a natural polymer. PLA is renewable, eco-friendly, sustainable and most importantly biodegradable [11]. Contradictorily, the 3D printing materials are themselves obstacles to the development of 3D printing technology due to some practical issues encountered during 3D printing. These are entirely unavoidable. Some of the 3D printing materials get a volume change and are subject to residual stress due to their crystalline nature, which could lead to dimensional problems. These are common with natural polymers, whereas synthetic polymers like ABS and PC are free of dimensional problems. However, synthetic polymers could be subjected to higher shrinkage. Synthetic polymers with higher melting and printing temperatures have a bigger temperature gradient and stress than natural polymers. Besides, the mechanical strength of natural polymers is less when compared with synthetic polymers. This could be due to simple linear molecular chains of natural polymer, as synthetic polymers are gifted with graft structures. But, natural-based 3D printing materials such as PLA have more tremendous advantages in comparison with high synthetic polymer materials such as ABS, nylon which include: (i) they do not pollute the environment when it gets degraded into H2O and CO2, whereas the synthetic raw materials come as petroleum by-product; (ii) no toxicity, hence they are widely applied for medical sutures, as drug sustained-release material, and for support in tissue engineering; (iii) no change in their plasticity and toughness for longer period. In the long-time service [12].

In a similar view, many efforts have been made to enlarge the list of biopolymer composite-based 3D printing materials in need of better performance. Some natural-based polymer hydrogels like collagen, alginate, chitosan and hyaluronic acid were preferred as 3D printing materials. These natural polymer hydrogels have gained great demand with their performance and especially with their potential for deployment in emerging areas like tissue engineering and regenerative medicine [5, 13, 14]. This has led to the evolution of bio-printed scaffolds loaded with cells.

The researchers show interest in optimizing the process parameters, such as raster angle, layer height, bed temperature, nozzle temperature, etc., to enhance interface bonding properties, mechanical properties, and accuracy. Also, the researchers have tried to modify the 3D material to improve mechanical properties by enhancing the cross-linking state and crystallinity of molecular chains. Incorporation with reinforcement has also been tried out in recent times. All these have resulted in making 3D printing materials with improvements in heat resistance, electrical conductivity, heat stability, fire retardance, and, most importantly, biocompatibility. Hence, the research has played a vital role in expanding the functional applications of FDM 3D-printed materials.

2. 3D Printing Technologies for Biomedical Prototypes

3D printing has become one of the dominating manufacturing techniques of the third industrial revolution, empowering and influencing our daily livelihood. Some of the additive manufacturing techniques are as follows [15]:

i. Selective laser sintering (SLS) involves metals to form 3D parts but requires post-processing like infiltration, sintering, and finishing to attain the finished part. Here, the metal powders are subjected to rolling across a build chamber and are directed into the powder by a laser or binder to form the part layer.

ii. Stereolithography (SLA), which uses a vat containing photopolymer resin and the resin activated by UV light to form layers,
iii. Inkjet printing, similar to SLA, deploys a movable inkjet head to apply a photopolymer. The photopolymer is then activated using a UV lamp,
iv. Laminated object manufacturing (LOM) utilizes large sheets of material to form. The individual sheets are rolled on the build platform, and then a binder is laid on the sheet. This is followed by rolling another sheet onto the binder, thus fusing the layers.
v. Fused deposition modelling (FDM) is a method of additive manufacturing that involves fabricating a part with a filament of polymer heated and deposited using an extruder nozzle.

Biomedical prototypes and biomedical structures are complex in their geometry; hence, it is challenging to fabricate them using conventional ways. The manufacturing difficulty could be addressed by their complex interior mechanical attributes and the material variations required. The need for prototypes for various purposes is to test for strength under given conditions so that it can be applied in real time. For example, prototypes could be tested for fractures, hair-line cracks, stresses, and damage to the bone could be investigated using prototypes [16, 17]. Recent research on orthopaedic devices is inclined towards developing replicate femurs using composite materials. Some of the advantages of these replicates are less inter-specimen variability, easy availability, and handling ease when compared with human bones. The initiation for realistic femur replicates was instigated in 1987 [18].

The 3D printing technique has the potential to fabricate biomedical prototypes and biomedical structures by controlling the process parameters locally. Some of these applications include implants, surgical devices, dental castings, orthotics, anthropology equipment and teaching aids. To fabricate medical prototypes with FDM processes, it is very important to thoroughly investigate the mechanical properties of the prototypes in order to establish constitutive models. This can then be related to the fabrication process parameters to change them locally during fabrication. The 3D filament gets melted in a heated nozzle and deposited in a layer-by-layer pattern over a platform. This depends on the deposition trajectory chosen, which significantly influences the behaviour of prototypes fabricated and their performance. Currently, there are three types of deposition methods [19]:

i. Rastering, where the path can be deposited in a single time, but the raster paths are usually inaccurate and of poor quality,
ii. contouring, where the boundary of each slice gets offset successively till the entire domain gets filled. This fits the boundaries of prototypes and is good quality. However, it yields material in distinct paths which require it to be deposited again, which slows the fabrication process.
iii. contour-raster combination, which is a combination of both contour and raster methods. The contouring takes care to fit the boundary, and the raster fills the deposition region on the inside. This ultimately results in good surface quality achieved relatively in a shorter period.

Besides all these, gap sizes between the adjacent paths can also influence the density of material deposition. The gap increases as the material density decreases. The filament orientation is another influencing factor on the mechanical performance of the prototypes.

3. Mechanical Properties on 3D Printed Dental Prototypes

The 3D printing technique plays a vital role in the fabrication of dental prototypes. The parts implanted for those who lost their teeth are called dental implants in biomedical applications. A dental implant is a surgical component that is placed into the jawbone to support a dental prosthesis, such as a crown, bridge, or denture. These dental implants are commonly made up of biocompatible materials such as polymers, titanium and titanium-based composites. Also, dental implants have common parts such as prostheses, abutments and implant post. Here, the dental post employs the artificial tooth root, which is placed into the jawbone via operation. The prosthesis is connected with the implant post with the help of an abutment. Also, the main visible part of the dental implant is the dental prosthesis, which may be a bridge or crown [20, 21]. These dental parts can be fabricated by the different 3D printing techniques illustrated in Figure 1.

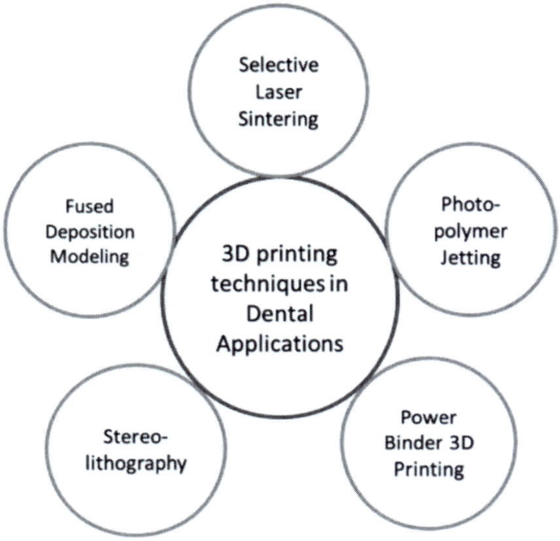

Figure 1. Diffèrent 3D printing techniques use in dental applications.

Dental implant is an emerging field of study in which the main aim is to develop the placement techniques, implant design and materials. Biocompatible materials are the only suitable materials for biomedical implants in specific applications. Moreover, the 3D printing method leads to the patient-specific design and unique implant structure, which helps dental implants integrate with surrounding tissues. Also, enhancing implant surfaces is the primary way to improve tissue integration with nanotextured surfaces. The major advantages of 3D printing in dental implants are that it is customizable, cost-effective and can be used for complex geometries. Also, it reduces the dental implantation surgical time and promotes more accuracy in the specific implant design for unique patients. The main limitations of 3D printing in dental implants are limited specific materials-based design, the requirement of additional testing before customization and limited accessibility [22].

Hence, the 3D printing for dental implants is based on mechanical strength, customization, precision, versatility, speed and material flexibility. However, the ability to create highly precise, patient-specific, and complex shapes and structures with 3D printing technology is revolutionizing dental implantology and opening up new possibilities for designing and fabricating dental implants, which offer improved performance, biocompatibility and long-term success rates.

The commonly used materials for fabricating dental implants are Titanium, Zirconia, Cobalt-chrome alloys, polymers and bioactive materials. These dental implant prototypes are usually applied in biomedical applications, surgical planning, design integration and fit testing [23].

This approach gives valuable assistance for the dental surgeon to fix the implant's position and choose the surgical method. Also, it can help explain the importance of dental implants, the development process, and the whole process to the patients before dental surgery [24]. Moreover, dental implants' precise shapes and structures can improve mechanical stability and performance. Based on the material selection for dental implants, titanium plays a vibrant role due to its higher strength, biocompatibility and high corrosion resistance. Unlike traditional methods, the main advantage of 3D-printed titanium-based dental implants is that they can make complex geometries. This can lead to better fit, comfort, and long-term outcomes for the patient.

Hence, zirconia is commonly used in 3D printed dental implants, a ceramic material with excellent mechanical and biocompatible properties for dental implant applications. This has many advantages compared to titanium-based dental implants, including enhanced aesthetics and reduced allergic conditions. Also, Cobalt-chrome alloys are utilized for fabricating the dental implants. Compared to Zirconia and Titanium, it has higher mechanical properties, which is the main reason for utilizing it in the fabrication of dental implants. The only drawback of these materials is that they are more challenging to work with than other materials in dental applications [25].

Other than titanium, zirconia and cobalt chrome alloys, the polymers, bioactive materials and ceramic alloys are used to fabricate the dental implant prototypes for the dental applications. Moreover, hydroxyapatite, which is a type of calcium phosphate, is a highly employed bioactive material for bone grafts, dental implants and bioactive glass. Over the decades, many parts have been replaced by 3D-printed implants in the human body. Dental implants should possess adequate mechanical properties due to their functional method.

Digitally, the lateral mandibular faults were generated to analyse the lateral mandibular defect caused by the disease to compare with the real-time analysis. The design of dental implants is done by 3-Matic and Mimics. The position of the implant is significant for the reconstruction operation, which can be planned by SKY Plan 4. Also, the high-precision 3D printer is used to fabricate and design the surgical guide. The stress distribution is analysed using the finite element analysis on the dental implant and surgical guide.

Hence, this solution can reduce the surgical time and preparatory things for the operation [26].

Also, the 3D-printed polymeric dental implant is another alternative for reconstructing the dental implant. The different sterilization approaches can alter the dental implants' mechanical properties, which are tested by the 3-point bending and tensile test via the universal testing machine. Conclusively, heat sterilization has improved the mechanical properties with temperatures of 121° and 131°C [27].

Moreover, the sintering temperature enhances the modulus and yield strength of 3D-printed titanium dental implants. The measurement of the porosity is playing a vital role in it. This method closes the void present in the implant structure, which is the main reason for having good mechanical properties with sintered- 3D printed titanium parts in dental applications [28]. Also, the thermocycling on the 3D-printed zirconia dental implants was studied for analysis of the mechanical properties. The thermal fatigue loading was applied on dental implants up to 5° to 55°C, which showed no improvement in the mechanical properties. Hence, thermal ageing at 90°C responded well and enhanced the mechanical properties [29].

On the other hand, the Digital Light Processing technique (DLP) is used to fabricate the 3D-printed zirconia dental implants. The mechanical properties of the different build angles were compared with the flexural strength of the Zirconia dental implants. Compared to these 0° and 45° build angles, the 0° build angle shows higher flexural strength [30].

Apart from designing the independent dental implants, patient-specific 3D printed parts are required to solve the dental issues. On this, 3D printed Poly-ether-ether-ketone (PEEK) dental implants were fabricated for the patient-specific conditions. The different speed of printing is also an important parameter. Here, it is shown that the PEEK dental implants with 30mm/s show excellent surface finish, higher compression and fatigue strength with the optimal processing parameters [31].

The future of 3D printed prototypes in dental implants is auspicious, but there are requirements in the improvements of high compressive mechanical strength, speed, accuracy and versatility of 3D printed dental implants. Developing unique anatomy dental implants needs to be tailored for patient-specific implants. Overall, the future of 3D-printed prototypes in dental implants is auspicious, which can be expected to see continued advancements and innovations in this field in the coming years. As the technology evolves, 3D-printed dental implants will become the standard of care for patients with missing or damaged teeth in dental applications.

Conclusion

In conclusion, 3D printing has emerged as a transformative force in many fields, including dental and biomedical applications. 3D printing can customize and precisely produce any complex structure that has revolutionized these fields. Moreover, ongoing research on heat sterilization and sintering has shown promise in improving the mechanical properties of 3D-printed biomedical and dental prototypes.

Overall, the 3D printing technique improves healthcare by introducing custom solutions for each individual. As technology improves, it can make healthcare more precise and individualized, leading to better treatment and healthier people worldwide.

References

[1] Perrot, A., D. Rangeard, and A. Pierre, "Structural built-up of cement-based materials used for 3D-printing extrusion techniques," *Materials and Structures/Materiaux et Constructions,* vol. 49, no. 4, pp. 1213–1220, 2016, doi: 10.1617/s11527-015-0571-0.

[2] Qaud, N., *Additive manufacturing technologies at Sulzer,* vol. 100, no. 2. 2018.

[3] Wall, M., *"Space Station's 3D Printer Makes Wrench From 'Beamed Up' Design,"* 2014.https://www.space.com/28095-3d-printer-space-station-ratchet-wrench.html.

[4] Anderson J, *"Full Circle: NASA to Demonstrate Refabricator to Recycle, Reuse, Repeat,"2017.* https://www.nasa.gov/mission_pages/centers/marshall/images/refabricator.html.

[5] Jungst, T., W. Smolan, K. Schacht, T. Scheibel, and J. Groll, *"Strategies and Molecular Design Criteria for 3D Printable Hydrogels,"* 2016, doi: 10.1021/acs.chemrev.5b00303.

[6] Kirchmajer, D. M., R. Gorkin, and M. In Het Panhuis, "An overview of the suitability of hydrogel-forming polymers for extrusion-based 3D-printing," *Journal of Materials Chemistry B,* vol. 3, no. 20, pp. 4105–4117, 2015, doi: 10.1039/c5tb00393h.

[7] Wang, X., M. Jiang, Z. Zhou, J. Gou, and D. Hui, "3D printing of polymer matrix composites: A review and prospective," *Compos B Eng,* vol. 110, pp. 442–458, 2017, doi: 10.1016/j.compositesb.2016.11.034.

[8] Ligon, S. C., R. Liska, J. Stampfl, M. Gurr, and R. Mülhaupt, "Polymers for 3D Printing and Customized Additive Manufacturing," *Chemical Reviews,* vol. 117, no. 15, pp. 10212–10290, 2017, doi: 10.1021/acs.chemrev.7b00074.

[9] Oskui, S. M. et al., *"Assessing and Reducing the Toxicity of 3D-Printed Parts,"* 2015, doi: 10.1021/acs.estlett.5b00249.

[10] Wang, Q., J. Sun, Q. Yao, C. Ji, J. Liu, and Q. Zhu, "3D printing with cellulose materials," *Cellulose,* vol. 25, no. 8, pp. 4275–4301, 2018, doi: 10.1007/s10570-018-1888-y.

[11] Rejeski, D., F. Zhao, and Y. *Huang, Research Needs and Recommendations on Environmental Implications of.* Elsevier B. V., 2017. doi: 10.1016/j.addma.2017.10.019.

[12] Liu, Z., Y. Wang, B. Wu, C. Cui, Y. Guo, and C. Yan, "A critical review of fused deposition modeling 3D printing technology in manufacturing polylactic acid parts," *International Journal of Advanced Manufacturing Technology,* vol. 102, no. 9–12, pp. 2877–2889, 2019, doi: 10.1007/s00170-019-03332-x.

[13] Himene, D. A. C., K. I. K. L. Ennox, R. O. R. K. Aunas, and A. K. K. G. Aharwar, *"Advanced Bioinks for 3D Printing: A Materials Science Perspective,"* 2016, doi: 10.1007/s10439-016-1638-y.

[14] Shen, X., J. L. Shamshina, P. Berton, and R. D. Rogers, *"Green Chemistry fabrication, properties, and applications,"* no. Mc, pp. 53–75, 2016, doi: 10.1039/c5gc02396c.

[15] Bhushan, B. and M. Caspers, "An overview of additive manufacturing (3D printing) for microfabrication," *Microsystem Technologies,* vol. 23, no. 4, pp. 1117–1124, 2017, doi: 10.1007/s00542-017-3342-8.

[16] Mc Donnell, P., M. A. K. Liebschner, W. Tawackoli, and P. E. Mc Hugh, "Vibrational testing of trabecular bone architectures using rapid prototype models," *Medical Engineering and Physics,* vol. 31, no. 1, pp. 108–115, 2009, doi: 10.1016/j.medengphy.2008.04.012.

[17] Mc Donnell, P., N. Harrison, S. Lohfeld, O. Kennedy, Y. Zhang, and P. E. Mc Hugh, "Investigation of the mechanical interaction of the trabecular core with an external shell using rapid prototype and finite element models," *Journal of the Mechanical Behavior of Biomedical Materials,* vol. 3, no. 1, pp. 63–76, 2010, doi: 10.1016/j.jmbbm.2009.03.002.

[18] Heiner, A. D. and T. D. Brown, "Structural properties of a new design of composite replicate femurs and tibias," *Journal of Biomechanics,* vol. 34, no. 6, pp. 773–781, 2001, doi: 10.1016/S0021-9290(01)00015-X.

[19] Li, L., Q. Sun, C. Bellehumeur, P. Gu, and P. Engineering, "Composite modeling and analysis of FDM prototypes for design and fabrication of functionally graded parts," *Solid Freeform Fabrication Proceedings,* no. January 2001, pp. 187–194, 2001.

[20] Khorsandi, D. et al., "3D and 4D printing in dentistry and maxillofacial surgery: Printing techniques, materials, and applications," *Acta Biomater,* vol. 122, pp. 26–49, 2021.

[21] Kim, J. C., M. Lee, and I.-S. L. Yeo, "Three interfaces of the dental implant system and their clinical effects on hard and soft tissues," *Mater Horiz,* vol. 9, no. 5, pp. 1387–1411, 2022.

[22] Saini, M., Y. Singh, P. Arora, V. Arora, and K. Jain, "Implant biomaterials: A comprehensive review," *World Journal of Clinical Cases: WJCC,* vol. 3, no. 1, p. 52, 2015.

[23] Sheela, U. B., P. G. Usha, M. M. Joseph, J. S. Melo, S. T. T. Nair, and A. Tripathi, "3D printing in dental implants," in *3D Printing in Medicine and Surgery,* Elsevier, 2021, pp. 83–104.

[24] Osman, R. B. and M. V. Swain, "A critical review of dental implant materials with an emphasis on titanium versus zirconia," *Materials,* vol. 8, no. 3, pp. 932–958, 2015.

[25] Liu, G. et al., "Development of bioimplants with 2D, 3D, and 4D additive manufacturing materials," *Engineering,* vol. 6, no. 11, pp. 1232–1243, 2020.

[26] Miljanovic, D., M. Seyedmahmoudian, B. Horan, and A. Stojcevski, "Novel and accurate 3D-Printed surgical guide for mandibular reconstruction with integrated dental implants," *Comput Biol Med,* vol. 151, p. 106327, 2022.

[27] Pop, S. I., M. Dudescu, S. G. Mihali, M. Păcurar, and D. C. Bratu, "Effects of disinfection and steam sterilization on the mechanical properties of 3D SLA-and DLP-printed surgical guides for orthodontic implant placement," *Polymers* (Basel), vol. 14, no. 10, p. 2107, 2022.

[28] Gagg, G., E. Ghassemieh, and F. E. Wiria, "Effects of sintering temperature on morphology and mechanical characteristics of 3D printed porous titanium used as dental implant," *Materials Science and Engineering: C,* vol. 33, no. 7, pp. 3858–3864, 2013.

[29] Zhang, F. et al., "3D printed zirconia dental implants with integrated directional surface pores combine mechanical strength with favorable osteoblast response," *Acta Biomater,* vol. 150, pp. 427–441, 2022.

[30] Osman, R. B., A. J. van der Veen, D. Huiberts, D. Wismeijer, and N. Alharbi, "3D-printing zirconia implants; a dream or a reality? An in-vitro study evaluating the dimensional accuracy, surface topography and mechanical properties of printed zirconia implant and discs," *J Mech Behav Biomed Mater,* vol. 75, pp. 521–528, 2017.

[31] Sonaye, S. Y. et al., "Patient-specific 3D printed Poly-ether-ether-ketone (PEEK) dental implant system," *J Mech Behav Biomed Mater,* vol. 136, p. 105510, 2022.

Editor's Biographies

Dr. Senthilkumar Krishnasamy is currently working as an Associate Professor at PSG Institute of Technology and Applied Research, Coimbatore, India. He graduated as a Bachelor in Mechanical Engineering from Anna University, Chennai in 2005. He then chose to continue his master studies and was graduated with Master degree in CAD/CAM from Anna University, Tirunelveli in 2009. He has obtained his PhD from Department of Mechanical Engineering – Kalasalingam Academy of Research and Education (2016). He has completed his postdoctoral fellowship at King Mongkut's University of Technology North Bangkok, North Bangkok and Universiti Putra Malaysia, Serdang, Selangor, Malaysia. His area of research interests includes modification and treatment of natural fibers, nano composites, hybrid reinforced polymer composites. He has published research papers in International journals, book chapters and conferences in the field of natural fibre composites. He is also editing books from different publishers.

Mr. Thitinun Ungtrakul holds both a Bachelor's degree (2007) and a Master's degree (2010) in Materials Science from King Mongkut's University of Technology North Bangkok (KMUTNB), Thailand. Following his graduation, he began his professional career as a Senior Analyst at Clariant (Thailand) Ltd., where he was responsible for material testing and analytical support in polymer applications. In 2015-present, he joined the Sirindhorn International Thai-German Graduate School of Engineering (TGGS) at KMUTNB as an Engineer in the Department of Materials and Production Engineering. In this role, he has been actively involved in academic research, technical training, and laboratory development.

In 2018-present, Mr. Ungtrakul has authored and co-authored several peer-reviewed journal articles and book chapters in the fields of materials science and polymer engineering. His research interests include polymer processing, polymer characterization techniques, polymer composites, and natural fiber-reinforced composites.

Dr. Sabarish Radoor received BSc (Polymer Chemistry) from Calicut University, Calicut, Kerala, India in the year 2006, MSc (Applied Chemistry) from Calicut University, Calicut, Kerala, India in the year 2008, MTech (Industrial Catalysis) from Cochin University of Science and Technology, Kochi in the year 2011 and PhD (Chemistry) from National Institute of Technology, Calicut in the year 2019. He has completed postdoctoral research at King Mongkut's University of Technology North Bangkok, Thailand, and Jeonbuk National University, Republic of Korea. He is currently working as a Postdoctoral Researcher at Myongji University, Republic of Korea. He has published numerous research articles in international journals, as well as book chapters and conference papers, primarily focusing on wastewater treatment and intelligent food packaging. He also serves as an editor for several books published by reputed international publishers.

Dr. Rajini Nagarajan is working as Professor in Department of Mechanical Engineering, Kalasalingam Academy of Research and Education (KARE), India. He has obtained his PhD from Department of Mechanical Engineering – Kalasalingam University (2013). Recently, he has received a Young Scientist Fellowship through the Start up Research Grant Project by DST, Govt of India. His research area is mainly focused on development and material characterization of hybrid composites. He has published more than 250 international peer reviewed journals and 30 international conferences in the field of natural fiber composites. Currently, he is guiding seven PhD scholars. He has completed his postdoctoral fellowship at Ghent University, Belgium in the field of Tribological Characteristics of plant fiber composites for medium load applications. He is also working as visiting Scientist in University Putra Malaysia in the field of hybrid composites.

Professor Dr.-Ing. habil. Suchart Siengchin is Advisor to the President of King Mongkut's University of Technology North Bangkok (KMUTNB), Thailand. He received his Dipl.-Ing. in Mechanical Engineering from University of Applied Sciences Giessen/Friedberg, Hessen, Germany, M.Sc. in Polymer Technology from University of Applied Sciences Aalen, Baden-Wuerttemberg, Germany, M.Sc. in Material Science at the Erlangen-Nürnberg University, Bayern, Germany, Doctor of Philosophy in Engineering (Dr.-Ing.) from Institute for Composite Materials, University of Kaiserslautern, Rheinland-Pfalz, Germany and Postdoctoral Research from School of Materials Engineering, Purdue University, USA. In 2016 he completed the Habilitation (Dr.-Ing. habil.) in Mechanical Engineering from Chemnitz

University of Technology, Saxony, Germany and worked as a Lecturer for Mechanical and Process Engineering Department at The Sirindhorn International Thai-German Graduate School of Engineering (TGGS), KMUTNB. He has been full Professor at KMUTNB and became the Vice President for Research and Academic Enhancement in 2012 and elected President of KMUTNB in November 2016 until January 2025. Currently he is Advisor to the President and Director of KMUB TechnoPark. He won the Outstanding Researcher Award in 2010, 2012, 2013 at KMUTNB and National Outstanding Researcher Award for the year 2021 in engineering and industrial research by National Research Council of Thailand (NRCT) and Honorary Fellow of RWTH Aachen university Award 2024. His research interests in Polymer Processing and Composite Material. He is Editor-in-Chief: Applied Science and Engineering Progress and International Advisory Board, eXPRESS Polymer Letters and Journal of Production Systems and Manufacturing Science and the author of more than 66 peer reviewed Journal Articles and edited books and book chapters more than 180 books. He has participated with presentations in more than 51 International and National Conferences with respect to Materials Science and Engineering topics.

Index

V

W

X

Z